AF719058

THE WISDOM OF ANCIENT COSMOLOGY

Also by Wolfgang Smith

The Quantum Enigma: Finding the Hidden Key

Cosmos and Transcendence: Breaking Through the Barrier of Scientistic Belief

Teilhardism and the New Religion

THE WISDOM OF ANCIENT COSMOLOGY

Contemporary Science in Light of Tradition

WOLFGANG SMITH

The Foundation for Traditional Studies

The Wisdom of Ancient Cosmology: Contemporary Science in Light of Tradition. Copyright © 2004 by Wolfgang Smith.

All rights reserved. Printed in the United States of America. No part of this book may be used or reproduced in any manner whatsoever without written permission except in the case of brief quotations embodied in critical articles and reviews.

For information address the Foundation for Traditional Studies, P.O. Box 370, Oakton, VA 22124.

http://www.traditional-studies.org
e-mail: ftsdc@erols.com

Library of Congress Control Number: 2003109192

ISBN: 0-9629984-7-8

Publication of this book was made possible by a grant from the Radius Foundation, and by contributions from Drs. J. C. Whitacre, II, James Stevenson, and David Bryce.

Cover illustration: The first, second, third, and fourth days of creation, from the Old Testament Miniatures in the Pierpont Morgan Library, MS M.638, f. 1. The images were probably the work of Parisian artists, from around 1250. Photo: Joseph Zehavy.

To Peter Schmitz-Hille, philanthropist and man of God,
in friendship and high esteem.

Contents

Foreword

That there are today, in our civilization, religions with followers still standing by their beliefs is, with respect to the modern world, a kind of anomaly: religious belief definitely belongs to a bygone age. A believer's situation, whatever his religion, is not an easy one then. But what is true for all sacred forms is especially true for Christianity, because for three centuries it has been directly confronted by the negations of modernity. The day when Hinduism, Buddhism or Islam experience the omnipresence of this modernity, they will undoubtedly in their turn undergo serious crises.

The blows dealt by the modern world against a people's religious soul is in the first place concerned with the plane of immediate and daily existence. No need for ideological struggle here; merely by the strength of its presence and extraordinary material success, this world refutes the world of religion, silences it, and destroys its power. This is because religion speaks of an invisible world, while contemporary civilization renders the sensory world more and more present, the invisible more and more absent.

This is, however, only the most apparent aspect of things. The omnipresence of a world ever more "worldly" is only the effect, in the practical order, of a more decisive cause that is theoretical in nature, namely the revolution of Galilean science, its technical progress being only its consequent confirmation. For the religious soul, the importance of the scientific revolution consists in the fact that it affects this soul's own inwardness. As powerful as it might be, for the human being, society represents only an environment which it can in principle ward off. Whereas the scientific revolution, insofar as it ascribes the truth to itself, imposes itself irresistibly and from within on the intelligence that it besieges. It is a cultural and therefore a "spiritual" revolution to the extent that it makes an appeal to our mind. But whenever it is a question of a believer's mind, it is the vision of the world and the reality implied by his faith that is subverted. What remains then is the option either to renounce his faith, or else—an almost desperate solution—to renounce entirely the cosmology that it entails.

On the whole Christian thought has committed itself to this second way: to keep the faith (but a "purified" faith!) and abandon all the cosmological representations by which that faith has been expressed. This is a desperate solution because these cosmological representations are first scriptural *presentations*, the very forms by which God speaks to us about Himself. But if we disregard these forms, what remains of our faith? Scripture informs us that the apostles saw Christ raised from the earth and disappear behind a cloud, while Galilean science objects that space is infinite, that it has neither high nor low, and that this ascension, even supposing it to be possible—which it is not—is meaningless. What remains is then to see in it a symbolic fiction by which the early Christian community attempted to speak its faith in a vanished Jesus Christ: if He is no longer visible, this is because He has "gone back to heaven." Following Rudolf Bultmann the majority of Protestant and Catholic exegetes and theologians have adopted this "solution." Since then an immense process of *demythologization* of Christian scriptures has been in progress. According to Bultmann, what is mythological is a belief in the objective reality of revelation's cosmological presentation: "descent," "resurrection," "ascension," etc. To demythologize is to understand that this cosmological presentation is, in reality, only a *symbolic* language, in other words, a fiction. To pass from myth to symbol, this is the hermeneutic that enables a modern believer, living at the same time in two incompatible universes—that of the Bible and of Galilean science—to avoid cultural schizophrenia.

But at what price? At the price of making unreal all biblical teachings on which faith relies and with which it is bound up. To reject this cosmological presentation, the witnesses of which the apostles, for example, vouch to have been, is this not to reject with the selfsame stroke the faith attached to it? What does this parting of faith from its cosmological garment, of kerygma from myth, imply? Basically, would this not separate the Divine Word from its carnal covering and ultimately deny the Incarnation?

How surprising that another way never occurred to Bultmann, a way which, had it been taken into consideration, might have changed many things in the course of the West's religious history. It is this way that the distinguished mathematician Wolfgang Smith proposes to explore, and into which he now offers us insights. In the present crisis, in which Christian thought is split between an impossible fideism and its confinement to moral problems, Wolfgang Smith's book discloses a liberating perspective which, in the name of science itself, restores to faith its entire truth. It would be hard to exaggerate the importance of such a work.

What is involved is a fact simple to state: the advent of the theory of relativity and of quantum physics entails the abandonment of the Galilean model of the universe, a definitive abandonment; or better put, the Galilean model remains valid as an initial rough estimate which gives a convenient—although quite inexact—description of the universe. However simple to enunciate, this fact is rife with extremely complex implications which science is still powerless to encompass within a general and unified theory. I will only say that our idea of the cosmos must be completely transformed both in its spatio-temporal structure (relativity) as well as in regard to the constitution of matter (quantum theory), assuming that it is still possible to make any representation of the universe at all. In any case, today it is nineteenth century materialism that has become a superstition. The rise of relativity and quantum theory, moreover, do not constitute recent events: they occurred at the beginning of the twentieth century. And so for a hundred years the cosmological and epistemological landscape of our culture has been reshaped profoundly. Philosophers, theologians, and exegetes are, however, far from realizing this; such is Bultmann's situation. The "scientific" vision of the world that he opposes to the mythological vision of faith is that of a science largely obsolete at the time when, in 1941, he expounded his program of demythologization (*Entmythologisierung*). And though he did occasionally bring up the theory of relativity, it was in passing, and in order to deny that the facts had significantly changed; thus he overtly continued to think within the spatio-temporal framework of nineteenth century materialist physics. And that has not altered. Still today the Catholic Church is mocked for condemning Galileo in the name of a retrograde worldview, whereas those who do so are themselves prisoners of an obsolete cosmology. One is ridiculed and blamed for belonging to such a Church, and made ashamed of a past judged disgraceful on grounds that have proved invalid.

It is high time to become truly aware of the cosmological revolution that has occurred. To this end I do not think there is a more useful and efficacious work than the one by Wolfgang Smith that I have the pleasure of prefacing. On the most essential points, the most burning questions concerned with biblical cosmology, heliocentrism, the nature of space and matter, the concept of a true causality, etc., he shows how the conclusions of contemporary science cease to be incompatible with the affirmations of traditional cosmology.

This does not, for all that, involve a new *concordism*, that is to say that more or less clumsy effort pursued in the nineteenth century (and even at the beginning of the twentieth) to reconcile the biblical with the scientific worldview. The error of this concordism lies not so much in seeking to rediscover the theses of physics in scriptural teachings (a need for unity is natural to reason), but in sharing with official science the conviction that the world is an exclusively material reality, entirely defined by its space-time coordinates. Basically, when it comes to creation, these theologians were as materialist as Descartes had been in regard to physics.

This concept is no longer supportable today, the notion of matter having been remarkably attenuated in contemporary physics. But, to understand this and provide a true interpretation, one requires recourse to the speculative keys offered by traditional cosmology. Wolfgang Smith accomplishes this task in a remarkably precise way, thereby revealing a depth of culture that is able to refer, for example, to the teachings of hermetic philosophy and Germanic theosophy (Jacob Boehme, Franz von Baader, etc.) in order to show what light Baader's concept of *intensive extension* sheds on the dogma of spiritual corporeity. I would add that this concept could also be related to that of *internal extension* (*extensio interna*), used in theology (especially by Suarez) to account for the real presence of Christ's *body* in the Eucharist: the latter is in fact to be found there with the distinction of its various parts and the relationships that order these parts among themselves, but according to an essential mode and unconnected to the exterior place that circumscribes them in its carnal mode of existence.

It is here, in the rigorous application of traditional keys, that the work of Wolfgang Smith manifests its originality and importance. As I have said, no concordism is involved. But neither is it a question of purely and simply rejecting relativity and quantum theory as having, by definition, no value before faith's transcendent affirmations. This negative attitude was that of René Guénon. Certainly, better than any other thinker, Guénon knew how to restore sacred metaphysics and cosmology in their truth, and this is rightly why Wolfgang Smith draws inspiration from his doctrine. But, with respect to recent physics, Guénon's attitude is somewhat misleading: for him it is only one production among others of a world that he condemns *en bloc*. Moreover, he does not seem to have any knowledge of quantum theory; as for relativity, he alludes to it but briefly and sees in it only a mathematical nominalism. This is quite regrettable, since with respect to infinitesimal calculus, a rather modern invention, Guénon knew how to demonstrate its importance and apply it (under its Leibnizian form) to solve some highly metaphysical questions.

Wolfgang Smith treats science differently. Without in any way aspiring to rehabilitate modern science, which quite often is only—as Guénon says—"*un savoir ignorant*," he yet observes that this science, by breaking away from the narrow materialism of classical physics, nullifies the objections raised by Galilean mechanics against the data of faith; at the same time, however, he imperiously demands that it be completed and rectified by a metaphysical interpretation based upon traditional doctrine.

I say *imperiously*, because the information provided by quantum physics is so paradoxical that it cannot be integrated into a realistic and intelligible theory built on modern conceptual resources alone. This was not the case with Galilean physics which seemed to have succeeded in reducing the world to a material expanse, by rendering physical reality and mathematical rationality identical, thereby eliminating any need for recourse to metaphysics. To Napoleon's question Laplace replied: "God? I have no need of this hypothesis!" Such is no longer the case with present day physics, and many among the greatest physicists, such as Niels Bohr or Werner Heisenberg, have been very much aware of this fact. Contrary to what Heidegger maintains, they have proven that science also strives to *think*. But to do so it is still necessary to wield conceptual instruments. I will not refer to the many questions broached by Wolfgang Smith, but only mention the admirable analyses developed in Chapter 7, "The Pitfall of Astrophysical Cosmology." First he sets forth the criticisms certain scientists have directed at the major dogma of the new cosmology, which is "big bang" theory: these criticisms reveal its weakness and even impossibility, and thus disqualify the use theologians have made of that theory. Next he establishes that physics, when applied to celestial bodies, having *de facto* no operational value in that domain, has necessarily an ontological significance, which however is illegitimate. If in fact sidereal bodies, as required by quantum theory, are composed of an almost nonexistent dust of particles, these bodies themselves, as identifiable realities, vanish into space. For a *body* (un *corps*) is also *a* body (*un* corps); a being that is not *a* being is no longer a *being*, says Leibniz. Now quantum theory has nothing to say about the existence of this unitary principle needed to account for the reality of a body: it is therefore truly incapable of accounting for the reality of any corporeal being, be it stellar or earthly (which is why some physicists have fallen into an idealism insupportable in other respects). Hence it is absolutely necessary, as Wolfgang Smith reminds us, to have recourse to what traditional philosophy calls a "substantial form," a unitary principle that endows a material body with its own reality. This is no speculative luxury that might be dispensed with, but a rigorously scientific need, since it is the incontestable truth of quantum physics itself that, for want of this substantial form, renders the reality of bodies forever *inexplicable* and indeed *impossible*.

We should be thankful to Professor Wolfgang Smith for having reminded us of these primary truths with the authority of a recognized scientist and the full resources of his broad philosophic and religious culture.

I also salute his courage, for he has dared to confront, with such constancy, the dominant ideology of modern culture, which is not without risk, to say the least. This ideology has turned science (a certain kind of science!) into the official mythology of our times. Basically, Wolfgang Smith shows us, with simplicity and sometimes with much humor, that Bultmann has chosen the wrong object: it is not religion but the customary interpretation of science that needs to be "demythologized." Only the doctrine of the *philosophia perennis* is able to accomplish this, and thereby to disclose the full truth of science itself.

Jean Borella
Nancy, France
June 3-10, 2002

Introduction

According to its title, this book has to do with ancient cosmology. Strictly speaking, however, its primary concern is with *traditional* cosmology, which is not the same thing. Yet the fact remains that the ancient cosmologies tended to be traditional in varying degrees, which means that the book has to do with ancient cosmology after all.

Nonetheless, it is imperative to distinguish between the two conceptions. To speak of "ancient" doctrine is to speak in historical terms; what renders a doctrine "traditional," on the other hand, is precisely the fact that it is *more* than historical, more than a mere historical contingency, which is to say that it embodies an element of revelation. What exactly that means, moreover, is something that only traditional doctrine itself can disclose. Suffice it to say that a doctrine is traditional by virtue of the fact that it partakes somewhat of eternity. It has power, therefore, to inspire the comprehensor, to awaken in us what the poet terms "intimations of immortality." Traditional cosmology, in particular, points thus beyond the cosmos; in the words of St. Paul, it leads from "the things that are made" to the "invisible things of God." Such a cosmology reconnects its votaries to the spiritual world; it is inherently religious, therefore, precisely in the original sense of *re-ligare*, "binding back."

The source of the aforesaid discrepancy between ancient and traditional cosmology can now be understood: it derives evidently from the collective human propensity to become forgetful of things spiritual, uncomprehending of the higher significations concealed, as it were, within doctrines of a traditional kind. On account of what St. Paul terms a "darkening of the heart," the spiritual content of sacred doctrine becomes progressively obscured. And this appears to be especially true in the case of cosmological teachings, the spiritual significance of which has become almost totally forgotten in modern times. This collective process of obscuration, moreover, traces back to the earliest historical periods, and was well under way even while the doctrines in question were still accepted as normative in their external sense. One knows, however, that "the letter killeth," and that the outer sense of a sacred doctrine cannot for too long survive the demise of its inner dimension. It is perhaps surprising, thus, that ancient cosmology survived in Europe, at least in some of its outer forms, for as long as it did: roughly until the Enlightenment, when it came to be replaced by paradigms of a very different kind.

One sees, in light of these observations, that one cannot expound or delineate traditional cosmology as one would, say, the facts of botany, or the history of Greece. Yet there are principles to which all traditional cosmology conforms, and these *can* be delineated, and can serve from the start as guide-posts along the path of discovery. I propose now to enunciate four such principles, in a kind of ascending order. The first is quite simple: it affirms that traditional cosmology has to do primarily with the qualitative aspects of cosmic reality, the very component, thus, which modern cosmology excludes. As we shall have occasion to see, this first recognition, simple though it be, has already enormous implications.

The second principle relates to the metaphysical notion of verticality and affirms a hierarchic order, in which the corporeal domain, as commonly understood, constitutes but the lowest tier. The transition from traditional to contemporary cosmology entails thus a drastic diminution, an ontological shrinkage of incalculable proportions, which of course pertains, not to the cosmos as such, but to the horizon of our worldview. As if to compensate for this reduction, contemporary cosmology imputes spatio-temporal magnitudes to the universe at large that stagger the imagination by their sheer quantitative immensity. The fact remains, however, that the spatio-temporal universe in its entirety constitutes but the outer shell, so to speak, of the integral cosmos, as conceived according to traditional cosmology.

The third principle presupposes the preceding two, and affirms that man constitutes a microcosm or "universe in miniature," which in a way recapitulates the order of the integral cosmos itself. This recognition, moreover, might well be singled out as the defining characteristic of the traditional worldview, which can in truth be characterized as anthropomorphic. I say "in truth," because what stands at issue is an ascription of anthropomorphism which is not merely poetical or imaginary, but factual. Tradition maintains that man and cosmos exemplify, so to speak, the same blueprint, the same master plan. This means, first of all, that even as man is trichotomous, consisting of *corpus, anima* and *spiritus*, so too does the cosmos prove to be tripartite, consisting of what Vedic tradition terms the *tribhuvana*, the "three worlds." Man, according to this view, is by no means a stranger in a hostile or indifferent universe, but constitutes the very heart and center of the cosmos in its entirety. I would like from the start to call attention to the fact that the possibility of human knowing is predicated upon this claim; as Goethe has beautifully put it: if the eye were not kindred to the Sun ("*wäre das Auge nicht sonnenhaft*"), it could not behold its light. In the final count, man is able to know the cosmos precisely because he is in fact a microcosm.

All this, needless to say, is diametrically opposed to the modern outlook. According to the prevailing worldview, man is indeed a stranger in the universe, an accidental and ephemeral product of blind forces. So far from constituting a microcosm, he is a most unlikely anomaly, a precarious molecular formation of astronomical improbability. Except for the laws of physics and chemistry, which are presumably operative in the cells of his body even as they are in stars and plasmas, he enjoys no kinship whatever with the universe at large, which presents itself as indifferent and ultimately hostile to his human aspirations. And again, let us note from the start that the new and ostensibly scientific cosmology is in principle incapable of accounting for the fact that man has the capacity to know, limited though his knowledge may be. It turns out (as we shall have occasion to see) that contemporary science is unable to account for even the most rudimentary act of cognitive sense perception, let alone for the higher modes of sensory and intellective knowing. The fact is that we can know the cosmos, because, in a profound sense which only authentic tradition can bring to light, the cosmos pre-exists in us. Only a cosmology, thus, which admits the traditional conception of man as microcosm, can account for what may well be termed the miracle of human knowing.

The fourth and final principle of traditional cosmology which I would like to point out pertains to its intimate connection with the spiritual ascent of man, as conceived according to the sapiential schools. It affirms that the higher strata of the integral cosmos can be known or entered experientially through the realization of the corresponding states of man himself, in what may indeed be termed an *itinerarium mentis in Deum*, a "journey into God," to employ St. Bonaventure's expressive phrase. The key to knowledge is thus to be found in the Delphic injunction "Know thyself": in the final count, there *is* no other way, no other means of knowing. We are able, at present, to know the corporeal world, because we have actualized the corresponding state: this is what constitutes, so to speak, our human condition. Occultists and New Age practitioners, it seems, are able in some instances to break into the lower reaches of the intermediary plane, sometimes termed the astral; to ascend beyond that level, on the other hand, is doubtless the prerogative of sages and saints. The relation between traditional cosmology and spiritual ascent is however twofold: not only are the higher levels of cosmic manifestation to be known through the actualization of corresponding spiritual states, but conversely, a certain preliminary knowledge of the cosmic hierarchy, as depicted, for instance, in iconographic representations, can serve as an aid to the spiritual ascent itself. From this point of view traditional cosmology becomes an adjunct to religion, a means to the attainment of spiritual realization.

Having thus introduced the subject of traditional cosmology, we need of course to ask ourselves whether that cosmology is compatible with the findings of contemporary science. Given what we know today about the universe—its origin, its configuration, and its laws—is it logically defensible to maintain the principles and tenets of that traditional cosmology? To be sure, most people today, be they scientists, philosophers, or theologians, would unhesitatingly answer in the negative. They take it as self-evident that modern science has once and for all disqualified the "primitive conjectures" of pre-modern cosmology. This judgment accords, moreover, with the prevailing evolutionist outlook, which perceives everything as arising "from below," and is therefore disposed to give pride of place to the latest turn of the evolutionary trajectory. One may wonder, of course, what the next turn might bring, and whether perhaps a still more highly evolved humanity could perceive things differently; but these are questions, in any case, which evolutionists are not prone to ask. Other individuals, comprising a less numerous category, profess high respect for the ancient doctrine, while they implicitly deny its truth. I am thinking especially of those who seem bent upon "psychologizing" every ancient cosmological belief, as if cosmology had to do simply with human fantasies. Suffice it to say that nothing could be more radically opposed to the traditional teachings, which invariably uphold the basic distinction between the human and the cosmic realms, and insist that cosmology refers indeed to the second of these domains; and one might add that the traditional conception of man as microcosm does not alter this fact, but indeed presupposes the distinction between the two realms. This brings us to a third group, which seems to take the ancient doctrines at their word while likewise accepting the outlook of contemporary science, as if there were not the slightest conflict or appearance of incompatibility between these respective claims. I have in mind, for example, individuals who cheerfully cast horoscopes and interpret these in more or less traditional terms, without realizing that this makes little sense in an Einsteinian universe.

Diverse as these respective mentalities may be, they exhibit a common deficiency. What I find conspicuously lacking in each case is any mark of critical acumen, any sign that a searching critique of the prevailing *Weltanschauung* has taken place; yet a critique that penetrates to the very foundations of that worldview is today the *sine qua non* for a sane approach to cosmology. Whatever we may think about the past, we live in a present dominated intellectually by the science of our day; and that science needs to be deeply probed and in a way transcended in order to access whatever treasures of wisdom the past may hold. As Theodore Roszak has sagaciously observed: "Science is our religion, because we cannot, most of us, with any living conviction, *see around it.*" It matters not whether we extol the wisdom of the past or cast horoscopes: so long as we do not "see around" science—or "break through the barrier of scientistic belief," as the subtitle of my first book puts it—we remain *intellectually* modern, profane, and indeed anti-traditional. This may be the reason, I surmise, why there are today few if any "intellectual" saints, the likes of St. Augustine, lets us say, or of St. Thomas Aquinas; it seems to be almost a precondition for sanctity, these days, to have escaped a university education, a fact which augurs ill for the state of theology. But be that as it may, my point is that "the barrier of scientistic belief" can indeed be breached, which is to say that it *is* possible, intellectually, to overcome the preconceptions of modernity and postmodernity alike, and thus to rejoin the pre-modern human family. This does not of course bestow instant illumination; yet what we do gain, most assuredly, by such an intellectual breakthrough, is a distant vision, at least, of high and sacred truths, which in itself is priceless and irreplaceable, and greater by far than any imagined wisdom which we do not possess. By the grace of God we do come to perceive truth, though it be "through a glass, darkly." We are indeed fortunate that the reception of wisdom is not an "all or nothing" proposition.

Implicit in all that I have said is the premise that the principles and tenets of traditional cosmology have not in fact been disqualified by the positive findings of contemporary science. To verify this contention, one needs to engage in the kind of critique alluded to above, and by way of a rigorous analysis, to arrive at a separation of scientific fact from scientistic fiction. Yet in reality this is only half of what needs to be done; for it is likewise imperative to interpret what science has disclosed, to make sense out of its positive findings, failing which one inevitably falls back into some kind of scientistic fantasy. I contend, however, that to arrive at an authentic interpretation of contemporary science one requires the resources of traditional doctrine itself. In an earlier monograph, entitled *The Quantum Enigma*, I have carried out an approach of this kind for physical science as such, with the result that its generic object—the physical universe, properly so called—could be integrated into the traditional ontologies as a sub-corporeal domain. As might be expected, this throws light on many questions, and explains things which hitherto had seemed incongruous or even paradoxical. As a rule, one finds that the discoveries of physics which strike us as the most bizarre are those that harbor a major metaphysical truth. Such is the case, for instance, with the well-known facts relating to Lorentz invariance—the phenomena, namely, of time dilation and Lorentz contraction—which admit of a metaphysical interpretation, as we shall have occasion to see in Chapter 5 of this book. Nothing that is true is lost by ascending to a traditional outlook: from a higher point of vantage one sees, not less, but more. The teachings of the traditional schools, so far from being disqualified by the discoveries of contemporary science, are in fact needed to arrive at a proper understanding of science itself.

Enough has been said, I presume, to indicate what this book is about. What I have to offer is not so much a set of answers to specific questions as it is a generic means of coping with what I perceive to be the major intellectual challenge of our time. The book exemplifies a methodical approach based upon traditional teachings, which in the end leads back, *Deo volente*, to the perennial wisdom of mankind. This is not to suggest that the task of interpreting science becomes a cut-and-dried affair once one has taken one's stand upon traditional ground: such is by no means the case. The enterprise presents its own challenge and has its own fascination, enhanced by the fact that the possibilities in this kind of inquiry are virtually endless. It is my hope that the few initial steps which I have taken along this path will encourage others to enter the field and continue these investigations along various lines; the collective need for the fruits of such labors is great. One might add, however, that having once arrived, by means of this approach, at the recognition of what contemporary science is actually about, one has no further need to conduct inquiries of this nature; one is then prepared intellectually to attend to "the one thing needful," which is the spiritual ascent itself.

One question remains: it may seem, from what has been said, that we are caught in a vicious circle. On the one hand, we need to "break through the barrier of scientistic belief" in order to gain access to the perennial wisdom of mankind, and on the other, we require the resources of that wisdom in order to "break through." Yes, such is indeed the case; as Christ Himself has declared: "Those who have, to them shall be given." What saves us in the present instance from outright paradox is the fact that knowledge or understanding of *sophia perennis* admits of degrees; as I have said before, we are fortunate that the acquisition of wisdom is not an "all-or-nothing" proposition.

This book, with the exception of Chapter 11, consists of previously published articles, written during the past few years, each under its own inspiration, so to speak. I did not originally intend that these articles should fit together as chapters of a book; but it happens that they do, due to the fact that each constitutes a step, if you will, along the path outlined above. These steps need not be conceived as successive, which is to say that the chapters can be read independently. The first two, let it be said, have been written for special occasions. Chapter 1, in particular, constitutes a slightly abridged version of my contribution to the Library of Living Philosophers volume in honor of Seyyed Hossein Nasr; and it is to be noted that Professor Nasr's response to that article (in conformity to the format of the LLP series) has been added to the present volume in the form of an appendix. As the reader might expect, the latter constitutes an invaluable commentary shedding light on the central issues. As concerns Chapter 2, it derives from a lecture delivered on two separate occasions in 1998: first as a "Templeton Lecture on Christianity and the Natural Sciences" given before a general audience at Gonzaga University, and later at a Thomistic symposium at the University of Notre Dame.

I am deeply grateful to Professor Nasr for suggesting the idea of the present anthology, and for offering, through the Foundation for Traditional Studies, to publish the book. It is a pleasure also to express my thanks to Katherine O'Brien, Executive Director of the Foundation, for her generous and expert help in piloting this project to completion. It remains only to express my hope that the book will prove to be of benefit to all who seek the truth.

Chapter I

Sophia Perennis and Modern Science

The relation of *sophia perennis* to natural science in the modern sense has been dealt with often and profoundly in the writings of Professor Seyyed Hossein Nasr. The considerations of the present chapter will take Professor Nasr's Gifford Lectures as their starting point.

The very first sentence presents what could well be termed their central thesis: "In the beginning Reality was at once being, knowledge, and bliss (the *sat, chit,* and *ânanda* of the Hindu tradition or *qudrah*, *hikmah*, and *rahmah* which are among the Names of Allah in Islam), and in that 'now' which is the ever-present 'in the beginning,' knowledge continues to possess a profound relation with that principial and primordial Reality which *is* the Sacred, and the source of all that is sacred."[1] An entire metaphysics, clearly, is alluded to and in a way implied by that opening statement; and that metaphysics, to be sure, is none other in essence than the *sanâtana dharma* of the Hindus, or what in the Western tradition has been named *philosophia priscorium* or *prisca theologia* (Marsiglio Ficino), *vera philosophia* (Gemistus Plethon), and *philosophia perennis* (Agostino Steuco) by turns.[2] However, given the anti-traditional bias of *modern* philosophy, not to mention the state of contemporary theology, the term "*sophia perennis*" will perhaps be the least misleading. The important thing to bear in mind is that this *sophia* or wisdom, when perceived from its own point of view, "is understood as the Sophia which has always been and will always be, and which is perpetuated by means of both transmission horizontally and renewal vertically through contact with that Reality that was 'in the beginning' and is here and now" (71), as Nasr explains.

It is a main point of the lectures that *sophia perennis* is intimately connected with "science" in a broad and distinctly pre-modern sense. "Sacred knowledge must also include a knowledge of the cosmos," Nasr maintains; and in fact, "one can speak of a *cosmologia perennis* which, in one sense, is the application, and in another, the complement of the *sophia perennis* which is concerned essentially with metaphysics" (190). One can say that every science, traditionally conceived, is an application of the perennial metaphysical wisdom by virtue of the fact that "all laws are reflections of the Divine Principle" (196), and a complement inasmuch as it constitutes *de jure* a support for the contemplation of the Principle itself. The traditional sciences, thus, are based upon the premise that the cosmos constitutes a theophany, and that, in the words of St. Paul, "the invisible things of Him from the creation of the world are clearly seen, being understood from the things that are made" (Romans 1:20). Science in the traditional sense is thus a matter of "reading the icon"—a far cry indeed from the Baconian vision! Science, as Bacon conceived of it, is concerned with the discovery of causal chains relating one phenomenon to another, an enterprise which can lead to prediction and control; traditional science, on the other hand, seeks to relate phenomena to the reality or principle of which they are a manifestation, an undertaking that leads ideally to enlightenment. In a word, the former is "horizontal" whereas the latter is "vertical" in its quest.

However, we must also take care not to make too much of this disparity; for it is to be noted that contemporary science at its best is not quite as Baconian as one might imagine on the basis of textbook lore. Think of Albert Einstein, for example, and his occasional remarks relating to "the Old One," suggesting that he too may have been searching for *vestigia* of a kind. It is on the level of epistemological presuppositions, in any case, that the distinction between the traditional and the modern conceptions of science assumes its sharpest form. We may not know what actually transpires in the mind of a contemporary scientist, but it is nonetheless clear what *ought* to transpire, according to the accepted canons: the scientist is supposed to reason upon data or information supplied by sense perception. It is all that he is officially permitted to do, if one may put it thus. The *sophia perennis*, on the other hand, provides for an incomparably greater range of cognitive possibilities, inasmuch as it maintains that the human intellect derives its "light" directly from the Divine Intellect: it "participates" in the Divine Intellect, as the Platonists say. All human knowing without exception hinges upon this "participation," which of course admits of various modes and countless degrees, ranging from the humblest act of sense perception to ways and intensities of knowing of which as yet we have not the slightest idea. But the fact remains: What ultimately connects the human subject to its object in the act of knowing is indeed "the true Light which lighteth every man that cometh into the world" (John 1:9).

There is however a fundamental difference between the knowing of an ordinary man and the knowing of an enlightened sage. Both may perceive a rock or a tree; but the one perceives it as a "thing," a self-existent entity—which in truth it is not!—whereas the other perceives it as a theophany, an entity whose essence and very being derive from the metacosmic Reality. It is the first kind of knowing, moreover, to which the Vedantic term "*mâyâ*" applies, for the world as perceived by the unenlightened *is* in a sense illusory: "For now we see through a glass, darkly" (1 Corinthians 13:12). It is however the contention of every sapiential tradition that this generic condition of nescience can be overcome, be it in full or in part, and that this rectification can indeed be effected in the present life through what Buddhists term "right doctrine" and "right method."

These are the things which we need to bear in mind in order to understand what *cosmologia perennis* is about. The fact is that every *bona fide* pre-modern science is rooted in an integral sapiential tradition, replete with a metaphysical doctrine and operative means, and requires moreover an ambience of this kind if it is not to wither and die, and thus give rise to what may indeed be termed a superstition.

An essential feature of the *cosmologia perennis* which will particularly concern us in the sequel is that it views the integral cosmos as a hierarchy of ontological degrees, what in Western tradition has sometimes been termed "the great chain of being,"[3] and used to be represented in Ptolemaic days by the so-called planetary spheres. One knows of course that Western man has abandoned the notion of "higher worlds" along with the Ptolemaic cosmography—the referent along with the symbol—and has opted instead for a *Weltanschauung* which would reduce the cosmos in its totality to what in fact constitutes, from a traditional point of view, its lowest plane: the domain of ponderable matter. This, I believe, is the decisive step that takes us into the modern world. One needs however to recognize that the reductionist hypothesis does not stand alone, but is mandated by what Nasr terms "the inherent limitations of the original epistemological premises of modern science" (206). These philosophic postulates, he maintains, plus the virtual disappearance in the West of the sapiential traditions, have prevented modern science "from becoming integrated into higher orders of knowledge, with tragic results for the human race" (207).

I consider this observation to be of capital importance, and singularly worthy of being pursued in depth. The object of the present chapter is to lay bare the offending epistemological premise and show how modern physics, freed from this impediment and duly reinterpreted, can indeed be "integrated into higher orders of knowledge" as Professor Nasr suggests.

As is well known, it was René Descartes who provided the philosophical basis of "classical" or pre-quantum physics by enunciating the distinction between *res cogitans* and *res extensa*. One generally perceives this Cartesian dichotomy as nothing more than the mind/body duality, forgetting that Descartes has not only distinguished between matter and mind, but has, at the same time, imposed a very peculiar and indeed problematic conception of the former element. He supposes, namely, that a *res extensa* is bereft of all sensible qualities, which obviously implies that it is imperceptible. The red apple which we do perceive must consequently be relegated to *res cogitans*; it has become a private phantasm, a mental as distinguished from a real entity. This postulate, moreover, demands another: one is now forced—on pain of radical subjectivism—to assume that the red apple, which is unreal, is causally related to a real apple, which however is not perceptible. What from a pre-Cartesian point of view was one object has now become two; as Whitehead puts it: "One is the conjecture, and the other is the dream."[4]

This, in a nutshell, is the fateful "bifurcation" hypothesis which underlies and in a way determines the *Weltanschauung* of modern science. The first thing, perhaps, that needs to be pointed out is that this Cartesian assumption can neither be proved by philosophical argument nor corroborated by scientific means. Whether it is indeed "tenable" is more difficult to say; however, bifurcation is in any case incompatible with the teachings of the traditional philosophic schools, not one of which has subjectivized the perceptual object in the manner of Descartes. According to the perennial consensus, we do "look out upon the world" in the act of perception, as every non-philosopher likewise believes; it is only that the world and the Reality are not exactly the same thing, which is however another question.

It is of interest to note that Whitehead attacks the idea of bifurcation on the ground that "Knowledge is ultimate."[5] What he means by this assertion is that the act of knowing cannot in principle be explained by reducing it to some natural process. And this position is traditional: "knowing" does not reduce to "being"; the two poles *chit* and *sat* are irreducible (and so is the third, the Vedantic *ânanda*, which however does not enter into our present considerations). Nonetheless, as Nasr points out: "In the beginning Reality was at once being, knowledge, and bliss . . ." Despite the irreducibility of "knowing" and "being" on the various planes of cosmic manifestation, the two are intimately related by virtue of the fact that *in divinis* "to know" and "to be" coincide.

Here, in this principial identity, lies, I believe, the ultimate explanation of what may well be termed the miracle of perception: the fact, namely, that in this quotidian act a subject and an object meet and in a sense become one, as Aristotle keenly observed. What we need above all to realize is that the cognitive union cannot in truth be consummated within the confines of the universe, which is and remains external to the human subject. There are light waves and sound waves, and there is brain function, to be sure; and these external or objective processes do no doubt play a necessary role. But they do not—they cannot!—constitute the perceptual act; to affirm that they do would be, once again, to reduce "knowing" to "being." The act itself, therefore, transcends perforce the bounds of space, and must by the same token be conceived as instantaneous or atemporal as well. The perceptual act, thus, is literally "not of this world." Is it any wonder, therefore, that post-medieval philosophy should have succumbed to the lure of "bifurcation"? Having lost sight of the Divine Intellect and denied in effect the mystery of "participation," is it surprising that post-medieval man should have implicitly denied the miracle of perception as well?

I will now take as my point of departure the following contention: What vitiates the customary interpretation of physics and prevents that science from being "integrated into higher orders of knowledge" is none other than the bifurcation postulate. This is the hidden premise one unfailingly assumes in the explication of scientific discovery. It is true that this postulate has been uncovered and attacked by some of the leading philosophers of our century—from Edmund Husserl to Alfred North Whitehead, Nicolai Hartmann, and Karl Jaspers, to mention but a few names—and yet that problematic tenet remains to this day unexamined and unopposed by men of science even in the sophisticated arena of the quantum debate, where just about everything else has been "put on the table." However, as I have shown elsewhere,[6] the premise can indeed be jettisoned, which is to say that nothing prevents us from interpreting physics on a non-bifurcationist basis.

Let us consider what this entails. It is clear, first of all, that to deny bifurcation is to give objective status once again to the perceptible things (red apples, for instance). Corporeal objects, let us call them. The first step, thus, in the proposed re-interpretation of physics may be characterized as the rediscovery of the corporeal world. This rediscovery or re-affirmation, however, does not constitute a return to a so-called "naive" realism, but demands a more refined and discerning ontology. We need in particular to take note of the following fundamental principle: "to be" is to be knowable. This is still realism, to be sure; clearly, it is the distinction between "knowable" and "known" that averts a lapse into idealism: the spurious reduction, that is, of "being" to "knowing." Evidently no such reduction is implied by the stated ontological principle. Every grain of sand in the universe is surely perceptible; but how many will ever be perceived? Now obviously the "naive" realist believes this as well, and one may ask why it should be necessary to abandon or to refine this common-sense position. What is the advantage, one might ask, of the proposed principle? What proves to be crucial is the following corollary: Different ways of knowing correspond to different kinds of being, or as we shall say, to different ontological domains. For example, corporeal being is the kind which can be known by way of sense perception. There are however other kinds of being which cannot be known by this particular means, and this is something a "naive" realism is ill-equipped to comprehend.

So much for the first step in the re-interpretation of physics; the second—as may now be surmised—is perforce the recognition of the physical as a separate ontological domain. Over the past centuries Western man has evolved a new and unprecedented way of knowing based upon measurement and artificial means of observation, which has brought to light a hitherto unrecognized category of objects: physical objects, we shall say. I have delineated the generic *modus operandi* of this cognitive enterprise in the previously mentioned monograph; suffice it to say that the observational process hinges upon an interaction between the *physical* object and a *corporeal* instrument, which then registers the result of the interaction by way of a perceptible state. The process thus renders "visible" in a sense what in fact is not, and thereby reveals a previously unknown ontological stratum. Our knowledge of this stratum has moreover progressed from the more or less crude approximations of classical physics to the incomparably more refined conceptions of quantum theory, which has revealed the physical to be in reality none other than the quantum world.

There are thus *two* ontological domains to be reckoned with: the corporeal and the physical; but the quantum theorist reckons only with one! On the strength of the bifurcation postulate he denies the corporeal, and thus in effect reduces the corporeal to the physical. The prevailing interpretation of physics has thus been vitiated from the start by a systematic confusion resulting from a failure to distinguish, in theory, between corporeal and physical objects. I say "in theory," because in practice everyone does evidently know the difference between a tangible scientific instrument, for example, or any other corporeal entity, and a cloud of quantum particles;

and that is of course the reason why physics has survived the confusion and is able to function. But the philosophy of physics does not fare as well. As Whitehead pointed out long ago in reference to the bifurcationist bias: "The result is a complete muddle in scientific thought, in philosophic cosmology, and in epistemology"; to which he adds: "But any doctrine which does not implicitly presuppose this point of view is assailed as unintelligible."[7] Let us hope that after seventy years of quantum debate there will now be a greater willingness on the part of scientists to consider a non-bifurcationist view of physics.

The non-bifurcationist interpretation has the immediate advantage of eliminating at one stroke what is generally called quantum paradox.[8] No need any longer for this or that *ad hoc* hypothesis to make things fit; no need for "parallel universes" or new laws of logic! The one thing needful to avoid the semblance of paradox is to jettison bifurcation once and for all.

What presently concerns us, however, is something else, which I consider to be more important still: the fact, namely, that physics, thus reinterpreted, can be "integrated into higher orders of knowledge," to avail ourselves once more of Professor Nasr's significant phrase. Let us consider how this integration comes about.

Modern physics, as I have said, has brought to light a hitherto unknown ontological stratum: the physical, namely, which in fact coincides with the quantum world. To be sure, this newly-discovered realm is nowhere referred to by the traditional schools, and it is open to question whether any ancient master could have divined the presence of such a domain. But though the physical stratum does not appear on the traditional ontological charts, it can be added: its position on the map can be ascertained. As I have shown,[9] the physical domain is situated between two traditionally defined levels: below the corporeal, namely, but above the so-called *materia secunda* that underlies the corporeal world.

Why, first of all, does the physical stand "below" the corporeal? The key idea has been supplied by Heisenberg—in *his* Gifford Lectures, no less!—when he pointed out that state vectors or so-called wave functions constitute "a quantitative version of the old concept of 'potentia' in Aristotelian philosophy," and referred to quantum objects as "a strange kind of physical entity just in the middle between possibility and reality."[10] In a word, the quantum level (and thus, in our view, the physical) stands to the corporeal as potency to act. But it happens that the principle of order in the hierarchy of ontological degrees may be conceived in the same Aristotelian terms: it is the vector from potency to act, if you will, that defines the ascending gradation. To say that the physical is in potency relative to the corporeal is therefore to situate the physical domain *below* the corporeal.

To proceed further, we need to remind ourselves that every traditional cosmology envisages one or more subcorporeal ontological strata. Among the twenty-five *tattvas* of the Sankhya, for example, it is *avyakta*, "the unmanifested," also termed *mûlaprakriti* or "root nature," that underlies the rest and thus constitutes the lowest stratum. It is evident, moreover, that the physical domain, made up as it is of things that can be discerned, is in act relative to *avyakta*, and is consequently situated "above" *avyakta*, above the absolute zero, so to speak, of the ontological scale. However, a less universal and thus a sharper and more enlightening "lower bound" to the physical can be found in the Scholastic tradition, which speaks of a *materia secunda* underlying the corporeal world.[11] We say "less universal," because this material substrate is not without determination, and is therefore distinguished from *prima materia*, the Scholastic equivalent of *avyakta*. What, then, is the nature of this determination? The answer to this literally most basic question concerning the corporeal domain has been given by St. Thomas Aquinas: the protomatter or material substrate of our world is said to be *signata quantitate*. Here is the key, I contend, to what physics is about.

There is no better way of explaining this connection than by the example of Euclidean geometry. Let us take the Euclidean plane and conceive of it in a pre-Cartesian manner, that is to say, not as a point set, but as a substrate or potency, in which neither points nor lines have yet been defined. One can say that points, lines, and indeed, all constructible figures, subsist potentially in that plane—until, that is, they have been actualized by way of geometric construction. One sees, however, that this plane also "carries" something else: a mathematical structure, namely, which we term "Euclidean" to distinguish it from other possible structures, such as the projective, the Lobachevskian, and so forth. Now, this structure manifests itself in certain geometric properties exhibited by constructed figures made up of points, lines and circles.[12] Let us observe, moreover, that whereas geometric figures are legion—there is an infinite number of them, as mathematicians are wont to say—the Euclidean properties are few, and fit together, so to speak, to constitute a coherent geometry, a single intelligible form; and that geometry or form, of course, is none other than the mathematical structure "carried" by the Euclidean plane.

We are now in a position to understand the rationale of physics from a traditional point of view. Replace the Euclidean plane by the aforesaid *materia secunda*, constructed geometric figures by physical objects, and the

Euclidean geometry by the "quantitative signature" of the *materia secunda*, conceived, once again, as a mathematical structure, and we have at hand the essential elements. What is particularly to be noted is that the objects of physics—its actual objects, that is, the kind that can affect a corporeal instrument or leave a track in a bubble chamber—are indeed "constructed," which is to say that they are defined or specified by a certain experimental intervention. This is the aspect of physics which led Eddington to stipulate that all fundamental laws can in principle be deduced on an *a priori* basis: look at the "net," he said, and you will know the "fish." Yes, up to a point. It is true, as Eddington has astutely observed, that the *modus operandi* of the experimental physicist affects the form of the physical laws or fundamental equations at which one arrives; but these laws or equations have also a content which does *not* derive from that *modus operandi*, even as the Euclidean properties of a constructed figure do not result from the process of geometric construction. The strategies of the geometer do of course affect the manifested geometric properties in the sense that a triangle and a circle, for example, exhibit the underlying Euclidean structure in different ways; and this exemplifies, once again, the subjective aspect of the scientific enterprise, which Eddington had his eye upon. But whereas the manifested geometric properties are indeed dependent upon the contingencies of geometric construction, they are nonetheless expressive of an objective mathematical structure, a given intelligible form in the Platonist sense. My point is that the laws of physics likewise manifest the mathematical structure of the *materia secunda* underlying the physical, and thus *a fortiori*, the corporeal domain.

I have conceived of the "*signata quantitate*," in light of contemporary physics, as a mathematical structure; but is this exactly what St. Thomas Aquinas had in mind? Whatever the Angelic Doctor may have been thinking of, it could hardly have been the Hilbert spaces and Lie groups of Hermitian operators with which contemporary physics is concerned. We must not, however, judge too hastily. What could be more strange, for example, than Plato's idea that "atoms" of earth, air, fire, and water correspond respectively to the cube, the octahedron, the tetrahedron, and the icosahedron. What indeed could be further removed from our contemporary scientific notions? And yet, as Heisenberg has brilliantly observed,[13] it turns out that Plato came as close to the quantum-theoretic conception of "elementary particles" as was possible in terms of mathematical structures available in pre-modern times. The point is, first of all, that the regular solids are "made of" polyhedra, and thus of entities which have no *corporeal* existence. One could say that Plato's "atoms" are mathematical as opposed to corporeal entities; and as such, they resemble the elementary particles of contemporary physics[14] and not the atoms of Democritus, or indeed, of pre-quantum physics. But why the regular solids of Euclidean geometry: what is special about these? What is special is that they are representations of a symmetry group; and so are the elementary particles of contemporary physics! It is only that the respective groups are different. This is not to suggest, of course, that Plato arrived at his conclusions by way of some rudimentary quantum field theory. He was doubtless looking at "atoms" from a very different point of view; and yet it could hardly have been an accident that he arrived at conceptions so strikingly similar in certain respects to our own.

The relevance of this example to the question of the "*signata quantitate*" is evident. Obviously St. Thomas, once again, is not looking at the problem from a point of view inspired by modern physics, and certainly one must be careful not to read such things as Hilbert spaces and Lie groups into an ancient text, the *Summa* no less than the *Timaeus*. But the crucial question is whether these mathematical structures are yet "quantitative" in an appropriate sense; and if that be the case, then the "*signata quantitate*" admits—by transposition, if need be—the interpretation which I have given above. As in the case of Plato's so-called atoms, it is at times necessary to look beneath the surface meaning of an ancient text to discern its contemporary relevance.

What, then, is the requisite conception of "quantity"? The answer, it appears, has been supplied by René Guénon when he observed that "quantity itself, to which they [the moderns] strive to reduce everything, when considered from their own special point of view, is no more than the 'residue' of an existence emptied of everything that constituted its essence."[15] Here we have it: Quantity is "the residue of an existence emptied of everything that constituted its essence."

It is clear, first of all, that cardinal number is quantity in the stipulated sense; after all, if we consider five apples, let us say, and take away their essence, what is left is no longer "five this" or "five that," but simply "five," the pure number. But it is to be noted that the notion of quantity, thus conceived, includes much else besides, and is more than broad enough to encompass the contemporary idea of mathematical structure. There is however something else that needs likewise to be pointed out: the tenet that the *materia secunda* of our world is indeed "*signata quantitate*" follows now from the very definition of "quantity" at which we have arrived. One could put it this way: What remains when all "content" has been evacuated from the universe must belong to the "container"; but that residue, by definition, is quantity.

We need to ask ourselves what it is that differentiates the physical universe from the material substrate; could it be "essences"? That position proves not to be tenable. We must remember that physical objects without exception are in a sense "constructed," which is to say that they are defined through a complex intentional process involving perforce an empirical intervention of some kind. Nothing is a physical object unless it has somehow interacted, directly or indirectly, with a corporeal entity. Physical objects, thus, are in a sense relational: they mediate between the material substrate and the corporeal plane. They have the "*esse*," if you will, of a potency waiting to be actualized in a corporeal entity; and thus, strictly speaking, they have no "essence," because they are not, in truth, a "thing." As Heisenberg has put it, they are "just in the middle between possibility and reality"; but only what is real partakes of essence.

It emerges that physics is *basic* but *inessential*; that is the crucial fact. It is basic because it descends to the material substrate, the *mûlaprakriti* or matrix of our world; but for that very reason it is inessential. The essence of a plant, after all, derives from the seed, and not from the ground in which the seed is planted.

It may seem paradoxical that a science whose ultimate object is the *materia secunda* should prove to be the most "exact" of all. Has not the subcorporeal been conceived traditionally as a primordial chaos from which the cosmos is brought forth by a determinative act, a divine command or *fiat lux*? Does not Genesis refer to this tenebrous realm as a *tohu-wa-bohu*, as "without form and void," and does not Proverbs speak of it as the *abyssos* upon which the divine Geometer "set His compass" to construct the world? We need however to recall that physics is concerned, not with *prima materia*, but with *materia secunda*, which is *signata quantitate*. The fact is that physics derives its exactitudes from "the white spot in the black field," to put it in yin-yang terms. We may rest assured that the mathematical structure of the material substrate has been inscribed by the great Geometer Himself; Dirac was not mistaken after all when he said that "God used beautiful mathematics in creating the world." One must not forget, however, that God used many other "beautiful things" besides mathematical structures, the point being that above the level of protomatter and of physical objects there are "essences" which likewise derive from the Divine Intellect. And these, to be sure, physics knows nothing about; the knowledge to which it gives access is "basic but inessential," as I have said.

Meanwhile the black field surrounding the white spot has likewise come into scientific view; on the fundamental level of quantum theory the physical domain has revealed itself as a partial chaos—to the consternation and chagrin of the "classical" physicist. The fact is that physical systems, quantum mechanically conceived, are in a superposition of states corresponding to the various possible values of their observables, even as the tone of a musical instrument is a superposition or composite of pure tones, each with its proper frequency. What is superposed in the quantum system, however, are not actual waves of some kind, but mere possibilities or *potentiae*, as Heisenberg says, which moreover are for the most part mutually incompatible. The quantum-mechanical description of a physical system depicts an ensemble of warring quasi-existences synthetically united; one wonders whether a more perfect characterization of semi-chaos could be conceived! I say "semi-chaos," because physical objects are evidently determined to some degree, on pain of having no objective existence at all; but this determination does not cancel the aforesaid superpositional indeterminacy, which remains as a witness, so to speak, to the primordial chaos that underlies our world. It appears that quantum mechanics has penetrated beneath the plane of *terra firma* to a depth approaching the level of the "waters" alluded to in Genesis, which remain in place even after "the Spirit of God" has moved over them. I find it remarkable how many major truths pertaining to the perennial cosmology have been unwittingly uncovered by twentieth-century physics, while scientists, for the most part, continue to view the traditional teachings as "pre-scientific superstitions."

Getting back to the quantum world, it is to be noted that the measurement of a dynamic variable turns out—once again, to the dismay of the classical physicist—to be an act of determination. Let us suppose that we are measuring the position of an electron. Prior to this measurement, this empirical intervention, the electron had presumably no position at all; it was most likely in a superposition of states corresponding to an infinite number of positions, continuously distributed over some appreciable and possibly vast region of space. And so it is until instruments put in place by the physicist interact with the electron so as to impose certain spatial bounds. The particle becomes thus confined, for a shorter or longer period of time, to a region of space small enough to count as a definite position. Now this scenario is disturbing, as I have said, to physicists accustomed to the pre-quantum outlook, which assumes that the physical object has a well-defined position, a well-defined momentum, and so forth, whether these quantitative attributes have been measured or not. But here again quantum theory stands on the side of the *cosmologia perennis*, which from time immemorial has viewed measurement as a determination, a creative act, like that of the geometer who constructs with the aid of his instruments.

Let us not fail to note that on a cosmic plane creative activity of whatever kind requires a pre-existent potency. If the divine Geometer had determined everything at one stroke, there would be nothing left for the human geometer to actualize; and as a matter of fact, the world as such could not exist. As every theologian knows, God alone is "fully in act"; which is to say that all other beings partake of potency in varying degrees. On every level, moreover, this potency or indetermination plays a most necessary and indeed beneficent role. According to St. Thomas Aquinas, even the human intellect is able to perform its cognitive function only by virtue of its radical potency, whereby it becomes receptive to whatever object presents itself, even as the emptiness of a container makes it receptive to all manner of concrete things. It must not be thought, therefore, that indetermination exists only in the quantum world; for indeed, it exists everywhere, on every ontological plane of the integral cosmos; and not, moreover, as a foreign element, a kind of blemish, if you will, but precisely as the natural complement of act. This is what the well-known figure of the yin-yang depicts with such grace, and that is doubtless the reason why Niels Bohr adopted this Taoist icon as his heraldic emblem. The notion of a cosmos made of *yang* (the "white" element) alone turns out to be unfounded and unrealistic in the extreme, and one wonders how this chimerical conception could have attained its strangle-hold upon the West; in any case, it was an insufficient physics that has for centuries confirmed us in this misbegotten notion, and it is a corrected and deepened physics which now apprises us of our long-standing blunder. Here again, on this fundamental cosmological issue, quantum theory sides with the traditional doctrine.

The decisive step in the restitution of the *cosmologia perennis* is without question the rediscovery of "forms" as an ontological and causal principle. Ever since Francis Bacon and René Descartes declared substantial forms to be a figment of the Scholastic imagination, Western science has labored to explain the whole in terms of its recognizable parts, or as one can also say: the greater in terms of the lesser. And not without success! As we know, the quest has led to the discovery of the physical realm, with its marvelous mathematical structures and undreamed of applications; it has not however led towards the realization of the reductionist goal. On the contrary, it turns out that the very discoveries of science point now in the opposite direction, as is evident if only one has eyes to see.[16] Meanwhile the reductionist philosophy appears also to have outlived its erstwhile usefulness as a heuristic principle. The scientist of the late twentieth century need hardly be motivated to investigate physical structures; instead, what he needs to realize, if further fundamental progress is to be achieved, is that there exist formal principles of a non-mathematical kind which also play a causal role, to say the least. These non-mathematical principles, to be sure, are none other than the aforesaid substantial forms, which prove moreover to be "essential" in a strict ontological sense. One should add that these forms or "essences" are mutually related and constitutive of a hierarchic order. What I have termed the rediscovery of the corporeal needs thus to be followed by the realization that this domain is itself stratified ontologically under the aegis of substantial forms.

The most obvious and important line of demarcation is given by the distinction between the organic and the inorganic, or better said, between living and non-living substances. One knows today that the distinction between the two realms is revealed with unprecedented clarity on the molecular level, where the difference between substances becomes in a sense quantified. In light of these findings it can now be said that the "distance" between the inorganic and the organic is of a magnitude that *de facto* rules out "accidental" transitions from the first to the second domain. At the present stage of scientific progress it is only on the basis of an unbending reductionist bias that this conclusion can still be denied.

A few words relating to the genetic code may be in order. Whether this magnificent discovery will serve to enlighten or further blind us depends, I believe, on the philosophical presuppositions which we bring to bear upon the issue. What we find in the DNA, clearly, is coded information, a coded message of incredible complexity; and this raises two questions. We need to ask ourselves, in the first place, what it is that has thus been encoded, thus expressed in a kind of molecular language; and we need to ask further by what means or agency this encoding has been effected. The reductionist, of course, assumes from the start that there is neither content nor agency beyond the molecular; but not everyone today agrees with this hypothesis. Robert Sokolowski, for example, has proposed that "it is the plant or animal form that encodes itself into the DNA," and that "the form is what the DNA serves to communicate."[17] There has been a growing recognition in recent years to the effect that the notions of substantial form and formal causation need once again to be taken seriously, not just by philosophers, but in the theory and practice of science as well. Among the benefits to science that can reasonably be expected from a sound ontology, the least, it would seem, is the reduction of futile research. To be more specific, such an ontology could dissuade scientists from searching for things that cannot possibly exist, such as, for example, the so-called "missing links" sought after by Darwinistically oriented anthropologists. By the same token, moreover, it could doubtless inspire life scientists to search for things that do exist but are out of range for a reductionist: "things in heaven and earth," namely, which are indeed not dreamed of in his philosophy. Most importantly, however, it should be clear from the outset that a living organism cannot be understood in depth without reference to the formal principle which constitutes its essence. Explanations "from below" have of course a certain validity and use; but their explanatory value is limited by the fact that they pertain, not to the living organism as such, but to mechanisms by which the organism fulfills its vital functions, which is not the same thing at all. Once again, one looks at the DNA but fails to recognize the plant or animal form which "encodes itself in the DNA," and which the DNA "serves to communicate."

We have alluded to the fact that the corporeal domain is stratified ontologically under the aegis of substantial forms; we should also remind ourselves, however, that according to the perennial doctrine the corporeal domain in its totality constitutes but the first and lowest tier of a larger cosmic hierarchy, consisting of three fundamental degrees.[18] What particularly concerns us is the fact that each level in this hierarchy comprises in a way all that exists below;[19] as Professor Nasr has put it: "Each higher world contains the principles of that which lies below it and lacks nothing of the lower reality" (199). This is a fact of immense importance; we need however to interpret the tenet with care, the point being that each higher level contains the *essential* principles of what lies below, and lacks nothing *essential* of the lower reality. What *is* however added in the passage from higher to lower states are certain conditions or bounds extraneous to essence, which in the case of the corporeal domain may be referred to summarily as *quantitative*, in conformity with our previous considerations. To put it as succinctly as possible: these constitutive factors of a quantitative kind are rooted in the *materia secunda*, revealed as *potentiae* on the physical level, and actualized on the corporeal. One finds thus that the mathematical structures displayed in the physical domain extend in a sense to the corporeal level,[20] *but not beyond*. What the physicist has his eye upon plays obviously a major role on the level of the perceptible world, but has no bearing whatsoever upon realities of a higher order; and even here below it perforce leaves out of account all that is essential, for the essence of corporeal things, as we have seen, is inexplicable in quantitative terms. To tell the truth, not even a perceptible grain of sand can be understood or explained in terms of physics alone—not to speak of living organisms, or the phenomenon of man.

Notes

1. The lectures have been published under the title *Knowledge and the Sacred.* The page numbers in the sequel are based upon the Crossroad edition of 1981.
2. See S. H. Nasr, op. cit., 69-71.
3. Arthur O. Lovejoy, *The Great Chain of Being* (Harvard University Press, 1964).
4. *The Concept of Nature* (Cambridge University Press, 1964), 30.
5. Ibid., 32.
6. *The Quantum Enigma* (Peru, Illinois: Sherwood Sugden, 1995).
7. *Nature and Life* (New York: Greenwood, 1968), 6.
8. *The Quantum Enigma* (cited in ref. 7), Chapter 3.
9. Ibid., Chapter 4.
10. *Physics and Philosophy* (New York: Harper & Row, 1962), 41. It needs to be pointed out that whereas Heisenberg relegates individual atoms and "small" atomic aggregates to the domain of potentia, he nonetheless regards "macroscopic" aggregates as actual entities, in keeping with the reductionist outlook. In light of non-bifurcation, on the other hand, one needs to distinguish ontologically between a corporeal entity X and the underlying atomic aggregate SX. The two are literally "worlds apart."
11. See especially René Guénon, *The Reign of Quantity* (London: Luzac, 1953), Chapter 2. The fact that the *materia secunda* underlies the physical domain as well will appear from the sequel.
12. These geometric properties are given in Euclid's axioms.
13. *Encounters with Einstein* (Princeton University Press, 1989), 83.
14. As Heisenberg has put it: "The 'thing-in-itself' is for the atomic physicist, if he uses this concept at all, finally a mathematical structure" (*Physics and Philosophy*, op. cit., 91).
15. Op. cit., 13.
16. It is to be noted that quantum mechanics, by its very formalism, puts an end to this reduction. A physical system, quantum mechanically conceived, is definitely not "the sum of its parts."
17. "Formal and material causality in science," *Proc. Amer. Cath. Phil. Assoc.* 69 (1995), 64.
18. The three degrees correspond to the *tribhuvana* or "three worlds" of the Vedic tradition, to *Beriah, Ietsirah* and *Asiah* of the *Qabbalah*, and microcosmically to the better-known triad *corpus-anima-spiritus*.

19. This ontological truth is symbolized in the Ptolemaic cosmography by the fact that the higher planetary spheres enclose the lower.

20. As I have explained in my monograph, there exists "a presentation-induced isomorphism between corporeal and sub-corporeal quantities" of which physicists make constant use. See op. cit., 80.

Chapter II

From Schrödinger's Cat to Thomistic Ontology

While the scientistic worldview continues to expand its grip on society, something quite unexpected and as yet largely unobserved has come to pass: this scientistic worldview, which still reigns as the official dogma of science, appears no longer to square with the scientific facts. What has happened in our century is that unprecedented discoveries at the frontiers of science seem no longer to accord with the accustomed *Weltanschauung*, with the result that these findings present the appearance of paradox. It seems that on its most fundamental level physics itself has disavowed the prevailing worldview. This science, therefore, can no longer be interpreted in the customary ontological terms; and so, as one quantum theorist has put it, physicists have, in a sense, "lost their grip on reality."[1] But this fact is known mainly to physicists, and has been referred to, not without cause, as "one of the best-kept secrets of science." It implies that physics has been in effect reduced to a positivistic discipline, or in Whitehead's words, to "a kind of mystic chant over an unintelligible universe."[2] Richard Feynman once remarked: "I think it is safe to say that no one understands quantum mechanics." To be sure, the incomprehension to which Feynman alludes refers to a *philosophic* plane; one understands the mathematics of quantum mechanics, but not the ontology. Broadly speaking, physicists have reacted to this impasse in three principal ways. The majority, perhaps, have found comfort in a basically pragmatic outlook, while some persist, to this day, in the attempt to fit the positive findings of quantum mechanics into the pre-quantum world-picture. The third category, which includes some of the most eminent names in physics, convinced that the pre-quantum ontology is now defunct, have cast about for new philosophic postulates, in the hope of arriving at a workable conception of physical reality. There seem to be a dozen or so worldviews presently competing for acceptance in the scientific community.

It is not my intention to introduce yet another *ad hoc* philosophy designed to resolve quantum paradox. I intend in fact to do the opposite: to show, namely, that there is absolutely no need for a new philosophic *Ansatz*, that the problem at hand can be resolved quite naturally on strictly traditional philosophic ground. What I propose to show, in particular, is that the quantum facts, divested of scientistic encrustations, can be readily integrated into a very ancient and venerable ontology: the Thomistic, namely, which as you know traces back to Aristotle. Rejected by Galileo and Descartes, and subsequently marginalized, this reputedly outmoded medieval speculation proves now to be capable of supplying the philosophic keys for which physicists have been groping since the advent of quantum theory.

First formulated in 1925, quantum mechanics has shaken the foundations of science. It appears as though physics, at long last, has broken through to its own fundamental level; it has discovered what I shall henceforth term the physical universe—a world that seems to defy some of our most basic conceptions. It is a world (if we may call it such) which can be neither perceived nor imagined, but only described in abstract mathematical terms. The most useful and widely accepted representation is the one formalized in 1932 by the Hungarian mathematician John von Neumann. In this model the state of a physical system is represented by a vector in a so-called complex Hilbert space. This means, in effect, that a state can be multiplied by a complex number, and that two states can be added, and that non-zero linear combinations of states, thus formed, will again be states of the physical system. Now, it is this fundamental fact, known as the superposition principle, that gives rise to quantum strangeness. Consider, for instance, a physical system consisting of a single particle, and then consider two states, in which the particle is situated, respectively, in two disjoint regions A and B, which can be as widely separated as we like. A linear combination of these two states with non-zero coefficients will then determine a third state, in which apparently the particle is situated, neither in A nor in B, but somehow in both regions. Now, one may say: State vectors actually describe, not the physical system as such, but our knowledge concerning the physical system. The third state vector, thus, simply signifies that so far as we know, the particle can be in A or in B, with a certain probability attached to each of the two possible events. A grave difficulty, however, remains; for the state of the physical system corresponding to the third state vector can in fact be produced experimentally, and when one does produce that state one obtains interference effects which could not be there if the particle were situated in A or in B. In some unimaginable way the particle seems thus to be actually in A and B at once.

What happens then if one measures or observes the position of the particle in the third state? It turns out that the act of measurement instantly throws the system into a new state. The detected particle, of course, is situated either in A or in B; which is to say that only unobserved particles can bilocate. All this, to be sure, is very strange; but let me emphasize that from a mathematical point of view all is well, and that in fact the theory functions magnificently. As I have said before, what puzzles physicists is not the mathematics, but the ontology.

Thus far I may have conveyed the impression that superposition states are rare and somehow exceptional. What is indeed exceptional, however, are states in which a given observable *does* have a precise value (the so-called eigenstates); and even in that case it happens that the system remains necessarily in a superposition state with respect to other observables. The quantum system, thus, is always in a state of superposition; or more precisely, it is at one and the same time in many different states of superposition, depending upon the observable one has in view. On the quantum level superposition is not the exception, but indeed the fundamental fact.

At this point one might say: There is no reason to be unduly perplexed; superposition applies, after all, to microsystems too minute to be observable without the aid of instruments. Why worry if "weird things" happen on the level of fundamental particles and atoms? Why expect that one can picture things or happenings which are by nature imperceptible? Most physicists, I believe, would be happy to adopt this position, if it were not for the fact that superposition tends to spread into the macroscopic domain. It is this quantum-mechanical fact that has been dramatized in the celebrated experiment proposed by Schrödinger, in which the disintegration of a radioactive nucleus triggers the execution of the now famous cat. According to quantum theory, the unobserved nucleus is in a superposition state, which is to say that its state vector is a linear combination of state vectors corresponding to the disintegrated and undisintegrated states. This superposition, moreover, is transmitted, by virtue of the experimental set-up, to the cat, which is consequently in a corresponding superposition state. In plain terms, the cat is both dead and alive. It remains, moreover, in this curious condition until an act of observation collapses its state vector, as the expression goes, and thus reduces it to one or the other classical state.

Of course, the mystery here has nothing especially to do with cats; it has to do with the role of measurement in the economy of quantum mechanics. Now, measurement is a procedure in which a given physical system is made to interact with an instrument, the resultant state of which then indicates the value of some observable associated with the system. For example, a particle is made to collide with a detector (a photographic plate, perhaps) which registers its position at the moment of impact. Prior to this interaction, the particle will in general be in a superposition state involving multiple positions; we must think of it as spread out over some region of space. Its evolution or movement, moreover, is governed by the so-called Schrödinger equation, which is linear, and hence preserves superposition, and is moreover strictly deterministic: an initial state uniquely determines the future states. At the moment of impact, however, this deterministic Schrödinger evolution is superseded by another quantum-mechanical law, a so-called projection, which singles out one of the positions represented in the given superposition state—apparently for no good reason!—and instantly assigns the particle to the chosen location. Now, this simple scenario exemplifies what happens generally in the act of measurement: a physical system interacts with an instrument or measuring apparatus, and this interaction causes the Schrödinger evolution of the system to be superseded by an apparently random projection. It is as though the trajectory of a particle, let us say, were suddenly altered without an assignable cause. Why does this happen? Inasmuch as the instrument is itself a physical system, one would expect that the combined system, obtained by including the instrument, should itself evolve in accordance with the corresponding Schrödinger equation; but in fact it does not! What is it, then, that distinguishes the kind of interaction we term measurement from other interactions between physical systems, in which Schrödinger evolution is *not* superseded?

Quantum theory holds many puzzles of this kind; the scandal of superposition assumes many forms. I would like to mention one more of these enigmas, which strikes me as particularly central. One might think of it as a simplified version of the Schrödinger cat paradox. In the words of Roger Penrose, the problem is this: "The rules are that *any* two states whatever, irrespective of how different from one another they may be, can coexist in any complex linear superposition. Indeed, any physical object, itself made out of individual particles, ought to be able to exist in such superpositions of spatially widely separated states, and so be 'in two places at once'! . . . Why, then, do we not experience macroscopic bodies, say cricket balls, or even people, having two completely different locations at once? This is a profound question, and present-day quantum theory does not really provide us with a satisfying answer."[3]

As you may know, these matters have been debated for a very long time, and various interpretations of the mathematical formalism have been proposed in an effort to make philosophic sense out of the theory. However, as Penrose observes, "These puzzles, in one guise or another, persist in *any* interpretation of quantum mechanics as the theory exists today."[4] After more than half a century of debate it appears that no clear resolution of the problem is yet in sight. One thing, however, one crucial point, has been consistently overlooked; and that is what I must now explain.

As one knows very well, it was the seventeenth-century philosopher René Descartes who laid the philosophic foundations of modern physics. Descartes conceived of the external or objective world as made up of so-called *res extensae*, extended things bereft of sensible qualities, which can be fully described in purely quantitative or mathematical terms. Besides *res extensae* he posits also *res cogitantes* or thinking entities, and it is to these that he consigned the sensible qualities, along with whatever else in the universe might be recalcitrant to mathematical definition. One generally regards this Cartesian partition of reality into *res extensae* and *res cogitantes* as simply an affirmation of the mind-body dichotomy, forgetting that it is much more than that; for not only has Descartes distinguished sharply between mind and body, but he has at the same time imposed an exceedingly strange and indeed problematic conception of corporeal nature, a conception, namely, that renders the external world unperceived and unperceivable. According to René Descartes, the red apple we perceive exists—not in the external world, as mankind had believed all along—but in the mind, the *res cogitans*; in short, it is a mental phantasm which we have naively mistaken for an external entity. Descartes admits, of course, that in normal sense perception the phantasm is causally related to an external object, a *res extensa*; but the fact remains that it is not the *res extensa*, but the phantasm that is actually perceived. What was previously conceived as a single object—and what in daily life is invariably regarded as such—has now been split in two; as Whitehead has put it: "Thus there would be two natures, one is the conjecture and the other is the dream."[5] Now, this splitting of the object into a "conjecture" and a "dream" is what Whitehead terms "bifurcation"; and this, it turns out, is the decisive philosophic postulate which underlies and determines our interpretation of physics. Beginning with his Tarner Lectures (delivered at Cambridge University in 1919), Whitehead has insistently pointed out and commented upon this fact. "The result," he declared, "is a complete muddle in scientific thought, in philosophic cosmology, and in epistemology. But any doctrine which does not implicitly presuppose this point of view is assailed as unintelligible."[6] I am here to tell you that today, after seventy years of quantum debate, the situation remains fundamentally unchanged. Just about every other article of philosophic belief, it would seem, has been put on the table and subjected to scrutiny, whereas bifurcation continues to be implicitly presupposed by physicists, as if it were a sacrosanct dogma revealed from on high. And so "the muddle in scientific thought" continues, and has only been exacerbated by the demands of quantum theory.

That's the bad news; the good news is that the situation can be remedied. In a recent monograph I have shown that physics can indeed be interpreted on a non-bifurcationist basis, with the result that quantum paradox disappears of its own accord.[7] No need any more for such things as the "many worlds" hypothesis or other *ad hoc* stipulations; to resolve the semblance of paradox one needs but to relinquish a certain philosophic postulate foisted upon us by Galileo and Descartes. Quantum paradox, it appears, is Nature's way of repudiating a spurious philosophy.

We need thus to take a second look at quantum mechanics, this time from a non-bifurcationist point of view. Now, to deny bifurcation is to affirm the objective reality of the perceived entity: the red apple, thus, is once again recognized as an external object. That perceptible entity, moreover, is to be distinguished from what may be called the "molecular apple," a thing, clearly, which can *not* be perceived, but can be known only through the methods of physics. One is consequently led to distinguish between two kinds of external objects: *corporeal* objects, which can be perceived, and *physical* objects, which can only be observed indirectly through the *modus operandi* of the experimental physicist. The two ontological domains are of course closely related, failing which there could be no science of the physical at all. The basic fact is this: Every corporeal object X is associated with a physical object SX from which it derives all of its *quantitative* attributes. The red apple, for example, derives its weight from the molecular. The crucial point, however, is that the two are not the same thing; X and SX belong to different ontological planes: to different worlds, one could almost say.

The bifurcationist, obviously, does not recognize this distinction, since he denies the existence of the corporeal object X; but in so doing, he implicitly identifies X with SX. The credo of bifurcation thus entails a reduction of the corporeal to the physical. And therein, in that reductionism, I say, lies the fundamental fallacy—the illusion, if you will—of the prevailing *Weltanschauung*.

Now, the amazing thing is this: whereas classical physics seemingly tolerates that error, quantum mechanics does not. It turns out that the new physics itself distinguishes between X and SX; it insists in fact upon that distinction—which is precisely what perplexes the physicist. In its very structure, that is to say, in its categorical distinction between the physical system and its observables, quantum mechanics affirms in its own way the ontological distinction between the physical and the corporeal planes. The system, thus, belongs to the physical domain, whereas the act of measurement terminates clearly on the corporeal, in the perceptible state, namely, of a corporeal instrument. It is true that the corporeal instrument I is associated with a physical system SI: but the point, once again, is that the two are by no means the same. What is special about measurement, thus, is the fact that it realizes an ontological transition from the physical to the corporeal domain. No wonder, therefore, that quantum theory should be conversant with two very different "laws of motion," for it has now become apparent that Schrödinger evolution operates *within* the physical domain, whereas projection has to do with a transit *out* of the physical and into corporeal. In the language of metaphysics one can say that the former describes a horizontal and the latter a vertical process. One can now see that the discontinuity of state vector collapse mirrors an *ontological* discontinuity; and that is the reason why the phenomenon cannot be understood from a reductionist point of view. State vector collapse is inexplicable on a physical basis because it results from the act of a *corporeal* entity.

These considerations strongly suggest that the superposition principle must be amended for subcorporeal systems, that is to say, for the SX of a corporeal object X; for it is altogether reasonable to suppose that the state vector of SX can admit only superpositions consistent with the perceivable properties of X. And that is no doubt the reason why cats cannot be both dead and alive, and why cricket balls do not bilocate. Penrose is absolutely right: if cats and cricket balls *were* "made of individual particles," they would indeed be able to exist in unrestricted states of superposition; but the point is that they are not thus made. From a non-bifurcationist point of view, corporeal objects, as we have seen, are not simply aggregates of particles, but something more. We need therefore to inquire what it is that differentiates X from SX; and for this we shall turn to the Thomistic ontology.

We must begin where St. Thomas himself began: namely, with the fundamental conceptions of Aristotle. The first step, if you will, in the analysis of being, is to distinguish between *substances* and *attributes*: between things that exist in themselves and things that exist in another. Having thus distinguished between what is primary and what is secondary, one proceeds to the analysis of the primary thing. The problem is to break substance into its components: to split the atom of substance, if you will; and for this one evidently requires the conception of things more primitive than substances, things "out of which" substances are made. Aristotle solved this problem with one of the great master-strokes in the history of philosophy: the distinction between *potency* and *act*. The customary definition of these terms is simple and quite unimpressive: That which is capable of being a certain thing, but is not that thing, is that thing in potency; whereas that which a thing already is is so in act. A seed is a tree in potency, and a tree is a tree in act. Aristotle goes on to define *matter*, or prime matter, to be exact, as that which is in potency to substance, to substantial being. Prime matter as such has consequently no being; but it has nonetheless a capacity or an aptitude for being, one can say. Now, what actualizes this capacity is indeed an act, and that act is called a *form*, or more precisely, a *substantial form*. Substance has thus been split into two components: into matter and form. It is the form, moreover, which contributes to the substance its essential content, its quiddity or "whatness," what the Germans so expressively call its *Sosein*. And yet that form is not itself the substance, is not itself the existent thing; for the form without matter does not exist.

It is at this point of the analysis that the genius of St. Thomas Aquinas becomes manifest. And here we come to a second master-stroke in the history of philosophy: St. Thomas recognized that substantial form is itself in potency to something else: to an act, namely, which is not a form; and that is the act-of-being itself. To put it in his own words: "The act-of-being is the most intimate element in anything, and the most profound element in all things, because it is like a form in regard to all that is in the thing."[8] Now, that innermost element constitutes the point of contact, as it were, between created being and its uncreated Source, which is God. The act-of-being, thus, belongs in the first place to God, who creates and sustains the universe; and yet it also belongs to created substance as its innermost reality. We may think of it as radiating outwards, through the substantial form, to the very accidents by which the being communicates itself to us.

Each being, moreover, is endowed with a certain efficacy, a certain power to act outside itself, which likewise derives from its act-of-being, and thus from God. And yet that efficacy, that power, is distinctly its own.

As Etienne Gilson has beautifully explained: "The universe, as represented by St. Thomas, is not a mass of inert bodies passively moved by a force which passes through them, but a collection of active beings each enjoying an efficacy delegated to it by God along with actual being. At the first beginning of a world like this, we have to place not so much a force being exercised as an infinite goodness communicated. Love is the unfathomable source of all causality."[9]

We are beginning, perhaps, to catch a glimpse of the Thomistic ontology; but let us continue. Not only is God's love the unfathomable source of all causality, but all causation, as we know it, *imitates* that love.

To quote Gilson once more: "Beneath each natural form lies hidden a desire to imitate by means of action the creative fecundity and pure actuality of God. This desire is quite unconscious in the domain of bodies; but it is that same straining towards God which, with intelligence and will, will blossom forth into human morality. Thus, if a physics of bodies exists, it is because there exists first a mystical theology of the divine life. The natural laws of motion, and its communication from being to being, imitate the primitive creative effusion from God. The efficacy of second causes is but the counterpart of His fecundity."[10]

This same Thomistic vision of Nature has been expressed by Meister Eckhart in a passage of rare beauty which I would like also to share with you. "You must understand," writes the German master, "that all creatures are by nature endeavoring to be like God. The heavens would not revolve unless they followed on the track of God or of his likeness. If God were not in all things, Nature would stop dead, not working and not wanting; for whether thou like it or no, whether thou know it or not, Nature fundamentally is seeking, though obscurely, and tending towards God. No man in his extremity of thirst but would refuse the proffered draught in which there was no God. Nature's quarry is not meat or drink nor clothes nor comfort nor any things at all wherein is naught of God, but covertly she seeks and ever more hotly she pursues the trail of God therein."[11]

Here we have it: a vision of Nature that penetrates to the very heart of things, to that "most profound element" which St. Thomas has identified as its act-of-being. And to be sure, this is no longer an Aristotelian, but an authentically Christian *Weltanschauung*. I propose to show next how the findings of quantum theory fit into that Christian worldview.

It needs to be pointed out, first of all, that the Thomistic philosophy, no less than the Aristotelian, is unequivocally non-bifurcationist. There is in either philosophy not the slightest trace of Cartesian doubt. What we know by way of sense perception are external objects, period; and these are the objects with which the Thomistic ontology is principally concerned. It follows that the findings of physics (in our sense) can be assimilated into the Thomistic worldview only on condition that they be interpreted in non-bifurcationist terms.

The fundamental problem, clearly, is to situate the physical domain in relation to the corporeal. Now, we know that transitions from the physical to the corporeal are effected by acts of measurement in which a certain possibility inherent in a given physical system is actualized; and this constitutes, Thomistically speaking, a passage from potency to act. Every physical system, in fact, is to be conceived as a potency in relation to the corporeal domain. I might add that this point has been made very forcefully by Heisenberg with reference to microphysical systems: "a strange kind of physical entity just in the middle between possibility and reality"[12] he called these, and went on to observe that in certain respects they are reminiscent of what he termed "Aristotelian *potentiae*." When it comes to the macroscopic domain, however, Heisenberg identifies in effect the corporeal object X with the associated physical object SX, and thus submits (as does virtually everyone else!) to a reductionist view of corporeal nature—as if the mere aggregation of atoms could effect a transition from potency to act. Non-bifurcation, on the other hand, implies, as we have seen, an ontological distinction between X and SX, which is to say that SX itself, no less than the quantum particles out of which it is composed, constitutes in fact "a strange kind of physical entity just in the middle between possibility and reality." To be precise, fundamental particles and their aggregates—be they ever so macroscopic!—occupy a position, ontologically speaking, between primary matter and the corporeal domain. Contemporary physics, it appears, has discovered an intermediary level of existence unknown and undreamt of in premodern times. It is this intermediary domain below the corporeal that I term the physical universe.

What then distinguishes the two ontological planes? From an Aristotelian or Thomistic point of view the answer is clear: what distinguishes a corporeal object X from SX is precisely its substantial form. It is this form that bestows upon X its corporeal nature and specific essence, its "whatness" or *Sosein*, as we have said. It is important to emphasize that this substantial form is *not* a mathematical structure; for indeed, if it were, X and SX would necessarily coincide. Substantial forms fall therefore beyond the ken of an exclusively quantitative science, a fact which Descartes himself clearly recognized. "We can easily conceive," he writes, "how the motion of one body can be caused by that of another, and diversified by the size, figure and situation of its parts, but we are wholly unable to conceive how these same things can produce something else of a nature entirely different from themselves, as for example, those substantial forms and real qualities which many philosophers suppose to be in bodies."[13] But is this not in fact the reason why Galileo and Descartes—protagonists of universal mechanism—rejected substantial forms, and banished sensible qualities from the external world? In so doing, however, they have cast out the very essence of corporeal being; one is left with a de-essentialized universe, a world emptied of reality.

We need today to free ourselves from the iron grip of this dehumanizing scientistic dogma. We need to rediscover the fullness of the corporeal world, replete with substantial forms and real qualities, and harboring deep within itself the mystery of what St. Thomas calls "the most profound element in all things." We have need of this discovery in every domain of life, including the scientific; but most especially, we have need of it in the spiritual domain. The fullness of the Christian life, in particular, demands a sacramental capacity on the part of matter which is totally inconceivable in terms of a reductionist ontology. There is no room for the Christic mysteries in a universe made up simply of fundamental particles. The deeper truths of religion have thus become unthinkable for us. In the final count, we know neither man nor the universe, because neither can be comprehended in separation from God; *I am the truth*, said Christ. To postulate, as we have, a self-existent universe productive of man, is to beget an illusion. Like the prisoners in Plato's cave, we are thenceforth confined to an illusory world, constrained to gaze upon a realm of shadows. Now, I surmise that of all the true philosophies—and I believe there are more than one—the Thomistic may be for us the safest and most efficacious means by which to effect the liberating intellectual rectification. Whosoever has sensed that "love is the unfathomable source of all causation" has already broken the chains; and whoever has grasped, even dimly, what St. Thomas terms the act-of-being, is well on his way.

Notes

1. Nick Herbert, *Quantum Reality* (Garden City: Doubleday, 1985), 15.
2. *Nature and Life* (New York: Greenwood Press, 1968), 10.
3. *The Emperor's New Mind* (Oxford University Press, 1989), 256.
4. Ibid., 296.
5. *The Concept of Nature* (Cambridge University Press, 1964), 30.
6. *Nature and Life*, 6.
7. *The Quantum Enigma* (Peru, Illinois: Sherwood Sugden, 1995). A helpful summary of the book with commentary has been given by William A. Wallace in "Thomism and the Quantum Enigma," *The Thomist*, vol. 61 (1997), 455-467.
8. *Summa Theologiae*, I, 8, 1.
9. *The Christian Philosophy of St. Thomas Aquinas* (University of Notre Dame Press, 1994), 183.
10. Ibid., 184.
11. *Meister Eckhart* (C. de B. Evans, trans., London: Watkins, 1924), vol. I, 115.
12. *Physics and Philosophy* (New York: Harper & Row, 1962), 41.
13. Cited in E. A. Burtt, *The Metaphysical Foundations of Modern Physical Science* (New York: Humanities Press, 1951), 112.

Chapter III

Eddington and the Primacy of the Corporeal

In his Tarner Lectures of 1938, published as *The Philosophy of Physical Science*, Sir Arthur Eddington has enunciated reflections on the nature of physics which to this day challenge our understanding of the physical universe. As is well known, Eddington championed a subjectivist interpretation of physical science: "selective subjectivism," he called it. Comparing the physicist to a fisherman catching fish with a net, Eddington contends that the known laws of physics can be deduced simply from the structure of the "net" itself. Even the basic numerical "constants of nature," as they are usually called, can be elicited, according to Eddington, by epistemological considerations alone. In a technical treatise entitled *Relativity Theory of Protons and Electrons* (Cambridge, 1936), he claims in fact to have accomplished such feats of mathematical derivation. One is bound to wonder, of course, whether these remarkable claims are justified: might there not be some gap or hidden flaw in these putative deductions? Suffice it to say that the significance and lasting relevance of Eddington's philosophy does not hinge on this question. It is no doubt true that his outlook, though widely respected, has rarely met with full belief: how could it be otherwise when an author, no matter how illustrious, purports to deduce—without recourse to a single experiment—that there are exactly 136×2^{256} electrons in the universe![1] I would however add that the more carefully one reads *The Philosophy of Physical Science*, the less fantastic such claims will seem.

Meanwhile the subjectivist interpretation of physics has in any case been gaining ground as a major trend, driven by the evolution of physics itself. In light of quantum theory, one has come to recognize that the act of measurement affects the system under observation in ways that can be neither predicted nor controlled. The presumed objectivity of physics has thus been compromised, and one hardly knows nowadays where to draw the line: how much of what the physicist "discovers" may be in fact the result of his own intervention? The very conception of what physics *is* has changed. As Heisenberg has put it: "Science [meaning physics] no longer stands before Nature as an onlooker, but recognizes itself as part of this interplay between Man and Nature."[2] This "interplay," moreover, is evidently effected by the act of measurement: it is here that the physicist acts upon Nature, and it is here too that Nature responds. Could it be, then, that Eddington was right: that the "net" of mensuration does determine the laws—and perhaps even the universal constants—of physics? On the strength of recent findings by an American physicist named Roy Frieden, this appears indeed to be the case. In a monograph entitled *Physics from Fisher Information*, published by Cambridge University Press in 1998, Frieden has essentially done what Eddington declared to be doable: from an analysis of the measuring process, Frieden has deduced the corresponding laws of physics. The analysis is information-theoretic, and Frieden employs a little-known information functional, discovered in 1925 by a statistician named Ronald Fisher. What takes place, according to Frieden, in the input space of a measuring instrument, is a transfer of Fisher information, from an information content J, "bound" to the phenomenon, to the acquired information content I from which the data are sampled. Frieden's great discovery is that the information difference I minus J (the so-called *physical* information K) satisfies a variational principle: roughly speaking, Nature contrives to minimize K.[3] It turns out that the physical information K constitutes in effect a universal Lagrangian, from which the laws of physics can be derived. The resultant approach—dubbed the method of extreme physical information: EPI for short[4]—has been successfully applied to the major domains, and no end of its scope is yet in sight. As a matter of fact, physicists seem for the most part to be interested in EPI, not so much on account of its enormous philosophical implications, but mainly because it is proving to be a powerful research tool. For instance, EPI is presently being applied to problems in quantum gravity and turbulence which have proved recalcitrant to other means of approach.

What concerns us is the fact that Frieden's spectacular results go a long way in support of Eddington's philosophy. Yes, it does appear to be the interaction between the measuring instrument and the measured system—the "net" and the "fish"—that accounts for the laws of physics, as the British savant had long ago foretold.[5]

I propose to present, first, a survey of Eddington's philosophy, followed by critical reflections consonant with the Aristotelian and Thomistic traditions. In particular, I shall argue that the very logic of Eddington's approach demands the distinction between what I have termed the physical and the corporeal domains, and in fact implies the ontological primacy of the corporeal.

Eddington begins his inquiry into the nature of physics with an appropriate definition of the physical universe: "Physical knowledge (as accepted and formulated today)," he writes, "has the form of a description of a world. We *define* the physical universe to be the world so described." (3)[6] The physical universe, thus, is not simply the universe as such, but the universe "so described": it is the world, one can say, as seen through the lenses of the physicist. Eddington contends that these "lenses" introduce subjective elements into our knowledge of the world, and thus into the physical universe, the universe "so described." The descriptions at which the physicist arrives are not wholly subjective, to be sure; but Eddington insists that the *laws* of physics, as distinguished from what he terms special facts, *are* wholly subjective. And that is the reason why these laws can indeed be known with mathematical precision: the vaunted precision of physics, says Eddington, derives in fact from its subjectivity. As Whitehead once put it: "Exactness is a fake." The special facts, on the other hand, are partly subjective and partly objective, "depending partly," as Eddington explains, "on our procedure in obtaining observational knowledge, and partly on what there is to observe." (66) Take for instance the fact that the Moon is so and so many miles from the center of the Earth: this fact evidently presupposes an observational procedure for measuring distance, but is not determined by that procedure alone. Not even Eddington would go that far! The aforesaid assertion concerning the Moon has thus an objective content; but for that very reason the distance in question turns out not to be precisely known. We will return to that side of the story later, in connection with quantum theory and the role of probabilities.

Having distinguished between laws and special facts, one finds the latter to be the more problematic, on account of their "partly objective, partly subjective" character. What about the structure of the atom as currently conceived? "When the late Lord Rutherford showed us the atomic nucleus," asks Eddington, "did he *find* it or did he *make* it?" (109) The question proves not to be well posed. If a hunter searches the forest and encounters a deer, he has clearly found or discovered the object of his quest. But what about the experimentalist who passes white light through a prism and observes the color green; has he "found" the color green (as the hunter has found the deer), or has he indeed "made" that color? Does the color green pre-exist in the white light, or has it been added as an effect of the experimental intervention? One could answer either way; and therefore, strictly speaking, the question admits of no definite answer, no simple "yes or no." It would certainly be misleading, in particular, to affirm that the color green was "found"—like a deer in the forest. The example of light diffraction, moreover, carries a very broad significance; as Eddington points out: "The question does not merely concern light waves, since in modern physics form, particularly wave form, is at the root of everything." (111) Getting back to the atomic nucleus, the case turns out to be indeed analogous to that of the color green: "It is equally unrigorous," Eddington contends, "to speak of the nucleus as having been 'discovered'. The discovery does not go beyond the waves which represent the knowledge we have of the nucleus." (111)

Continuing his inquiry into the nature of the physicist's "net," Eddington distinguishes between the application of scientific instruments and the conceptual frame of reference in terms of which empirical data are interpreted. "Whatever we have to apprehend," he notes, "must be apprehended in a way for which our intellectual equipment has made provisions." (115) And in a passage which reads like a page out of Immanuel Kant, he goes on to explain:

> The epistemological method of investigation leads us to study the nature of the frame of thought, and so be forewarned of its impress on the knowledge that will be forced into it. We may foresee *a priori* certain characteristic which any knowledge contained in the frame will have, simply because it is contained in the frame. These characteristic will be discovered *a posteriori* by physicists who employ that frame of thought, when they come to examine the knowledge they have forced into it. Procrustes again![7] (116)

What, then, does Eddington have to tell about this all-important "frame of thought"? "For a scientific outlook," he writes, "I think the most fundamental of all forms of thought is the *concept of analysis*. This means the concept of a whole as divisible into parts, such that the co-existence of the parts constitutes the existence of the whole." (118)

A word of comment is called for at this point. In light of the Aristotelian definition of *quantity* as that which admits mutually external parts, one sees that acceptance of the so-called concept of analysis has in principle restricted the purview of physics to the quantitative domain. And this fact itself explains why "the concept of substance has disappeared from fundamental physics" (110): it has disappeared because the concept of substance is not a quantitative notion. Substance, as the Scholastics taught, derives from essence or substantial form, whereas "*Numerus stat ex parte materiae.*" The concept of analysis, therefore, serves to filter out, not only qualities, but likewise substantial forms. When Eddington tells us that "what we ultimately come down to is *form,*" it is a very special kind of "form" he is alluding to: mathematical form, one can say, the equivalent of mathematical structure.

But let us return to Eddington's train of thought: he now goes on to point out that the concept of analysis is further narrowed by what he terms the atomic concept, or the concept of identical structural units. "The new conception," he explains, "is not merely that the whole is analysable into a complete set of parts, but that it is analysable into parts which resemble one another . . . I will go farther, and say that the aim of the analysis employed in physics is to resolve the universe into structural units which are *precisely* like one another." (122)

This too calls for a word of comment. The concept of analysis, in its weaker version, eliminates substantial form from the whole, but not necessarily from the parts. However, the additional requirement that the parts shall be "precisely alike" does in effect de-substantialize the parts as well. It entails that mathematical structure is all that remains—a goal, I would add, which cannot be fully realized, for the simple reason that in the absence of all substance *nothing whatever remains.* Yet even so the goal proves to be useful as a heuristic principle. Situated, as it is, in the physicist's fundamental frame of thought, the idea, presumably, has consequences. Meanwhile, as everyone knows, "fundamental particles" have indeed been found (or "made"!), and while these are not all alike, they do break into classes within which the members *are* "precisely like one another." Why, then, has this occurred? As Eddington goes on the explain:

> Granting that the elementary units found in our analysis of the universe are precisely alike intrinsically, the question remains whether this is because we have to do with an objective universe built of such units, or whether it is because our form of thought is such as to recognize only systems of analysis which shall yield parts precisely like one another. Our previous discussion has committed us to the latter as the true explanation. We have claimed to be able to determine by *a priori* reasoning the properties of the elementary particles recognized in physics—properties confirmed by observation. (123)

One is perhaps beginning to see that Eddington's claim to have deduced the so-called cosmical number $136x2^{256}$ may not be quite so absurd after all.

It is safe to say that so far few scientists have sided with Eddington when it comes to his subjectivist interpretation of particles. It should however be pointed out that the findings of Roy Frieden, to which I have already alluded, do lend considerable support to that contention as well. The first indications in that regard can already be found in Frieden's monograph, where it is shown that there is indeed a connection between the information-theoretic quantities I and J, and the physics of corresponding particles. For example, it turns out that "the intrinsic information I in the four-position of a free-field particle is proportional to the square of its relativistic energy mc^2," and that "the constancy of the particle rest mass m and Planck's constant h is implied by the constancy of J."[8] The fact that propositions of particle theory can thus be translated into the language of information theory does indeed suggest that we may be dealing ultimately with information-theoretic conceptions. Meanwhile research in the applications of EPI to particle theory is yielding results; as Roy Frieden put it during a recent phone conversation: "We are now getting particles!" But how could one "get particles" from an analysis of the measuring process unless it is the measuring process itself that gives rise to the particles? Never before have Eddington's subjectivist contentions appeared more plausible from the standpoint of physics.

The theoretical analysis of things physical into particulate parts goes hand in hand with corresponding experimental means designed to effect the desired decomposition. As Eddington puts it:

> In examining microscopic phenomena, we have to bear in mind the Procrustean methods of the experimenter which contrive to supply what our frame of thought demands. Like the sculptor, he renders visible the parts or combinations of parts which our analytical imagination creates; or at least his sorting and manufacturing operations produce effects which humour our belief that the parts are there. (131)

The logic of Eddington's thought leads inescapably to the final conclusion: "The fundamental laws of physics are simply a mathematical formulation of the qualities of the parts into which our analysis has divided the universe; and it has been our contention that they are all imposed by the human mind in this way, and are therefore wholly subjective." (131) Only I would add that the term "qualities" as applied to the parts seems inappropriate and should be replaced by the term "relations." Physics, then, has to do with the relations between parts which its own analysis has introduced.

There are those who think that the mathematical structures discovered by the physicist are simply there—somewhat as Mt. Everest is said to be; but Eddington evidently disagrees:

> Theoretical physics is highly mathematical. Where does the mathematics come from? I cannot accept Jeans' view that mathematical conceptions appear in physics because it deals with a universe created by a Pure Mathematician: my opinion of pure mathematicians, though respectful, is not so exalted as that. (137)

Eddington's point is this: "The mathematics is not there till we put it there." We begin to put it there, moreover, the moment we define the most basic of all measures: the measures of spatial distance and temporal duration. According to Eddington's analysis, this is accomplished by way of the mathematical theory of groups, which thus stands at the base of all physics. It is by means of mathematical groups constituted by what Eddington calls "terminable sets of operations" (140) that mathematics gains a foothold, so to speak, on the ground of external reality.[9] Or better said: on the ground of knowledge; for as we have noted, it is Eddington's contention that the mathematics pertains, not to the world as such, but to our *knowledge* of the world. It is thus subjective: we ourselves have put it there. And this means that the physical universe—the universe "so described"—is likewise subjective in precisely the same sense: it is subjective because it is constituted by mathematical structures which the physicist himself has imposed. This is Eddington's central point, and ultimately his *only* point: the conclusion to which all his arguments lead.

But let us continue: Starting with terminable sets of operations, one arrives at the abstract mathematical group in which the original operations are represented by symbols, such as the familiar x, y, and z of algebra, which now function as undefined elements of the group. What counts—and this is indeed the crux of the matter—are not these symbols, but the mathematical structure represented by their means: it is this structure, and this alone, that has physical significance. As Eddington clearly explains:

> Physical science consists of purely structural knowledge, so that we know only the structure of the universe which it describes. This is not a conjecture as to the nature of physical knowledge; it is precisely what physical knowledge as formulated in present-day theory states itself to be. In fundamental investigations the conception of group-structure appears quite explicitly as the

> starting point; and nowhere in the subsequent development do we admit material not derived from group-structure. The fact that structural knowledge can be detached from knowledge of entities forming the structure gets over the difficulty of understanding how it is possible to conceive a knowledge of anything which is not part of our own minds. So long as the knowledge is confined to assertions of structure, it is not tied down to any particular realm of content. (142)

It may seem at first glance that in the preceding passage Eddington has simply enunciated the Aristotelian and Thomistic doctrine of knowledge: it is by way of a form that we know an external object. But one must remember that the knowledge of which Aristotle and Aquinas speak is *objective*: we know the very form that resides in the object (be it a substantial or an accidental form). Why, then, is the knowledge of mathematical forms or structures said to be subjective? That seems to contradict Eddington's contention that it is precisely by way of mathematical forms that the physicist is able "to get out of his mind" and make contact with the external world. There is however no contradiction here. Yes, by way of the mathematical form the physicist gains knowledge of the external world; Eddington's point, however, is that the form in question has been artificially imposed: "The mathematics is not there until we put it there." And it is for this reason, and in this sense, that our knowledge of mathematical structures—our physical knowledge of the world—is said to be subjective.

The so-called physical universe—"the world so described"—turns be out to constituted by mathematical structures which we ourselves have imposed; in a word, it proves to be man-made. Yet this way of putting it is also misleading; for inasmuch as physical knowledge is partly objective, "the world so described" must be "partly objective" as well. One is left with a curiously equivocal conception, which may indeed enlighten the wise but is bound to deceive the unwary.

One thing, at least, is certain: If there be a "man-made" world, there must also be a world which is *not* man-made. And to be sure, Eddington himself alludes occasionally to that other world: "I have referred also to an objective universe," he writes, "which cannot be identified with the universe of which the above-mentioned knowledge [the knowledge of physics] forms a description." (158) But what exactly does Eddington have to tell about that other world? What is the nature of that universe which stands "behind" the physical? The crucial question is this: How do we *know* that universe? And indeed: Do we know it at all? Physics tells us how that world interacts with the strategies of the physicist: it tells us the "answers" it gives to our physical questions; but is that all we know?

It will be instructive to recall the conception of Eddington's "two tables," of which he tells in the Introduction to an earlier book[10]: "I have settled down to the task of writing these lectures and have drawn up my chairs to my two tables. Two tables! Yes; there are duplicates of every object about me—two tables, two chairs, two pens." And he proceeds to explain that the first of the two tables is the perceptible, the one with which we are familiar in daily life, whereas the second is the molecular table, which we do not perceive. At first glance it may seem that Eddington is distinguishing between what I have termed the *corporeal* domain and the physical, and that his "other universe" may in fact be the corporeal: the world in which the perceivable table is contained. But this proves not to be the case. The surmise is negated (some 260 pages later) when Eddington has this to say on the subject of sense perception:

> When messages relating to a table are travelling in the nerves, the nerve-disturbance does not in the least resemble either the external table that originates the mental impression or the conception of the table that arises in consciousness.[11]

He goes on to allude to the complex process by which the "nerve-disturbance" is transformed into the perceived table, a process accomplished "partly by instinctive image-building inherited from the experience of our ancestors, partly by scientific comparison and reasoning." He concludes that "by this very indirect and hypothetical inference all our supposed acquaintance with and our theories of a world outside us have been built up."

Here we have it at last: the old Cartesian philosophy! The "first table," it turns out, has been relegated to the nebulous realm of consciousness; as we should have surmised from the start, Eddington is bifurcationist in his *Weltanschauung*. The "external table" has thus to be inferred from phantasms conjured up somehow in response to nerve messages to which the former gives rise. But what concerns us at the moment is the objective world; or better said: our knowledge of the objective world. Assuming that the aforesaid leap of inference can be accomplished—a big assumption, to be sure!—the question is whether it lands us in the objective universe, or in the physical. And given that it is by way of mathematical structure that we "get out of our mind," the answer is clear: what we arrive at can be none other than "table number two": the molecular table belonging to the *physical* world. On the basis of Eddington's philosophical postulates, there is no way out of the impasse: the objective universe remains unknown and unknowable.

But then, how does Eddington know that there *is* an objective world, a world "beyond" the physical? Eddington is suspicious of ontological claims: "It is an advantage of the epistemological approach that the question of attributing a mysterious property called 'existence' does not arise," (156) he tells us. What he means is that in speaking of things that we know, the term "existence" ceases to be "a mysterious property" and becomes simply a way of saying that we know the things in question. "I myself," he explains, "often say that atoms and electrons exist. I mean, of course, that they exist—or are—in the physical universe, that being the theme of discussion." (156) Well and good; yet the fact remains that in alluding to an objective world—a world which, according to his own premises, we do not and cannot know—Eddington has evidently forfeited what he terms "the advantage of the epistemological approach": he has left the comparatively safe ground of phenomenology and has ventured onto metaphysical turf. Unwittingly, it seems, and against his stated intentions, he has become a metaphysician. And in my opinion, a metaphysician of the worst kind: a *Cartesian* metaphysician, that is.

Eddington was of course cognizant of the fact that philosophers held opinions at variance with his own. He was aware, in particular, that L. Susan Stebbing, in a book entitled *Philosophy and the Physicists*, had taken aim at the position he espoused. "Some of the pure philosophers," he notes, "deny that the scientific description applies to the objects which in ordinary speech are called physical objects.[12] Their opinion is voiced by Prof. Stebbing: 'He [the physicist] has never been concerned with *chairs*, and it lies beyond his competence to inform us that the chairs we sit upon are abstract'." (159) Eddington does of course perceive the enormous implications of Stebbing's contention: "If the physicist is not concerned with chairs," he goes on to say, "the astrophysicist is not concerned with stars. There is one Professor of Astrophysics, Prof. Dingle, who has not been afraid to recognize this logical conclusion." What, then, is Eddington's response? "Prof. Dingle, like Prof. Stebbing," he tells us, "has dropped the outlook which determines the familiar use of words, and has strayed into a world where men look at things in the way philosophers would have them look, and language is diverted to describe these things which philosophers consider most worth attention." (160) But here, quite clearly, Eddington has it exactly backwards: it is the bifurcationist—the individual who takes such a familiar object as a chair to be a phantasm residing in our brain or consciousness—that has "dropped the outlook which determines the familiar use of words"; it is he that has "strayed into a world where men look at things in the way philosophers would have us look": the way *Cartesian* philosophers would have us look, to be precise. Ask anyone who happens not to be a modern-day intellectual! I need not belabor the point: it is far too obvious. What is truly strange and indeed problematic, on the other hand, is that Eddington himself should have missed the point: that he seems not to have noticed that in ordinary speech a chair is precisely a familiar object of perception and *not* an aggregate of quantum particles or a Schrödinger wave function in Hilbert space. I find it surprising that he seems not to perceive the incongruity of sitting on an aggregate of quantum particles, especially after we have been told that these aggregates are "partly subjective": how can one sit on a "partly subjective" chair? And for that matter, how can one sit on a "mathematical structure"? My colleagues in mathematics would find this hard to comprehend. What is missing, of course, in mathematical structures, is *substance*: the very thing that has been "filtered out" by the physicist. A chair without substance, it turns out, cannot be sat upon.

Clearly, in responding to his philosophic critics, Eddington has entered a realm far removed from the domain of his own expertise. Gone are the lucidity and quasi-mathematical precision of his former discourse, when he lectured us on the nature of the physical universe, following what he termed the method of epistemological analysis. It was here, in this domain, that he arrived at those major insights, the validity of which physicists are only now beginning to confirm. So long as Eddington remains faithful to his epistemological approach, or to what philosophers call the way of phenomenology, all is well; but his doctrine becomes muddled the moment bifurcation comes into play. Understandably so: phenomenology and bifurcation do not marry. It was Edmund Husserl, after all, the father of phenomenology, who fired the first salvo against bifurcation, and in so doing awakened the philosophic world from its Cartesian slumber. Yet to this day the scientific community has paid little heed. With very few exceptions, scientists remain committed, at the most fundamental level of their thought, to the Cartesian philosophy, which they have imbibed at a tender age, and have been too busy, ever after, to question or investigate.

My point is that bifurcation constitutes a foreign element in Eddington's philosophy. The postulate is extraneous to the central insights of that philosophy, and functions mainly as a source of confusion and inconsistency. I would go so far as to say that bifurcation goes against Eddington's philosophic instinct and better judgment, and that it could assert itself in his thought only because it was never detected as the unwarranted postulate it is. If only Eddington had been able to apply his epistemological approach to the problem of human perception, his views with regard to the "first table" would have been vastly different. He would then have doubtless discovered that here too "the concept of analysis" has come into play, with the result that one has covertly replaced the living organism by an artificial model, a mere aggregate of parts, bereft of life and consciousness. But his opinions on that subject were obviously arrived at by other means, and when he speaks of nerve impulses, image formation, and the like, he is simply repeating the conventional wisdom. It was James J. Gibson who did in fact apply the "epistemological method" to the problem of human perception, and in so doing has laid that conventional wisdom to rest.[13] Eddington, I suspect, would have been delighted, had he lived to see the day. He would, I think, have recognized the American as a kindred spirit: another hard-headed scientist with little patience for humbug. But let us not forget: it took Gibson decades of painstaking research—much of it experimental—to arrive at his epoch-making conclusions; and obviously, Eddington was otherwise occupied.

Strictly speaking, Eddington *has* no worldview: his postulates preclude that the world can be viewed at all. The problem lies in his refusal to admit human sense perception as a means of obtaining a more-than-physical knowledge of the external world. Nothing would be more natural, I say, and indeed more in keeping with the logic of Eddington's thought, than to suppose that the objective world houses such familiar things as Professor Stebbing's chairs; but as we have seen, at this juncture Eddington turns Cartesian.

He need not have done so. The whole purpose of the Cartesian axioms was to replace the perceptible realm by the physical: that imagined mechanical world, namely, made up of so-called *res extensae.* It was for this that the familiar objects had to be exiled, banned to the limbo of consciousness. But as one knows today, that postulated mechanical world has proved to be chimerical: physicists no longer believe in a clockwork universe. The realization has dawned that the physical universe—the world of the mathematical physicist—is indeed "partly subjective"; and it was Eddington himself who refined and systematized that recognition, better perhaps than anyone else. Where is the necessity, then, let alone the justification, for the Cartesian postulates? It was belief in a mechanical universe that dethroned the perceptible world some three centuries ago;

but today the picture has changed. With the demise of Cartesian *res extensae*, the door has been opened to the restitution of the perceptible realm. Having at last awakened from his Galilean slumber, the scientist is free once more to accept what mankind has believed all along: the fact, namely, that we do look out onto the world.

But there is more to be said on this question: I shall argue that the very logic of Eddington's approach calls for the aforesaid restitution of the perceptible realm. It is by way of sense perception, after all, that we know the scientific instruments which constitute the concrete or non-mental component of Eddington's "net." And it is by means of this "net," let us recall, that we define and impose the mathematical structures that constitute the physical domain. Does it make sense, then, to maintain that the concrete instruments of the experimentalist are *physical*? Strictly speaking, it does not: a case indeed of putting the cart before the horse. The physical belongs, after all, not to the objective world, but to "the universe so described"; as Eddington says, it is "partly subjective." What is needed, however, are *objective* instruments! In other words, one requires instruments, not "instruments so described." I contend that one can no more measure with physical instruments than one can sit on a wave function.

What saves the day for physics, of course, is the fact that the experimentalist does not accept the Cartesian philosophy; which is to say that he treats his apparatus, not as a mathematical structure, but as a perceivable object. As there are said to be no atheists in the trenches, so one finds that there are no bifurcationists in the laboratory. All knowledge of the external world begins in the perceptible realm: deny the perceptible object, and nothing external remains. Deny, in particular, the scientific apparatus which all can perceive, and you have denied Eddington's "chisel" and "net": the very center of his doctrine has then been negated.

Having spoken of sense perception as the primary means of knowing the external world, it now behooves us to reflect upon that contention. How *do* we "get out of our mind" to perceive an external object? Eddington informs us that the physicist gets out of *his* mind by way of mathematical forms: "The fact that structural knowledge can be detached from knowledge of entities forming the structure gets over the difficulty of understanding how it is possible to conceive a knowledge of anything which is not part of our own mind." Something analogous can now be said of sense perception: we likewise perceive the external object by way of forms; it is only that the forms in question are not mathematical. If they were, we would perceive, not tables and chairs, but mathematical structures. No one, moreover, has ever perceived a mathematical structure: as has long been recognized, one arrives at mathematical structures, not by direct perception, but by a process of abstraction.

How the forms by means of which we perceive are transmitted from the perceived object to the percipient is a question with which we need not concern ourselves; what matters is that these forms *are* transmitted, and that they constitute the connecting link between the knower and the known. We know the very forms that subsist in the object, the very forms, in fact, that *constitute* the object. However, we must not interpret this doctrine too simplistically; it does not mean that we know the object "without residue." On the contrary: In the very act of knowing, we know the object to be "more" than what is given, more than what we are able cognitively to possess. In a word, we perceive the object as a *transcendent* entity. The object is transcendent, moreover, not simply because it has an existence of its own, but because it conceals within itself an immensity, an unfathomable depth. It is of course in the poet, the mystic, and the authentic philosopher, that these recognitions come to the fore; and yet it appears that some inkling, at least, of that transcendence, of that hidden immensity, is implicit as an essential component of all human perception—so long as that perception *is* human.

So too, in defining the corporeal domain to be the perceptible, it needs likewise to be understood that the corporeal is in a sense *more* than the perceptible. The human eye, for example, does not see light above or below a certain range of frequencies, nor does it perceive objects that are exceedingly large or small, and in any case it sees only the external surface. And yet it is by way of sense perception, limited though it be, that we know the corporeal world, to the extent that we do know it. One can say that the corporeal universe is the world to which human sense perception gives access. And this indeed is *our* world: the world in which we find ourselves. This corporeal universe, moreover, is in fact the only *objective* world which our human faculties—sensory *and* mental—allow us to know. It is true that we do also have knowledge of the physical universe, which is categorically imperceptible and hence not corporeal; but as Eddington shows, that physical universe is not objective. The corporeal universe, one finds, constitutes indeed the one and only objective world we know, and sense perception constitutes evidently our one and only means of knowing that objective world.

Having noted that the perceived object is perceived to be transcendent, I wish now to point out that it is precisely by virtue of this perceived transcendence that the given entity is known to be objective. Transcendence, one can say, constitutes the mark or criterion of objectivity.[14] And let us observe, by way of contrast, that in the case of a physical object, rigorously conceived, no such transcendence is discerned. Such an object is simply the mathematical structure it is defined to be; as Eddington explains: "It has come to be the accepted practice in introducing new physical quantities that they shall be regarded as *defined* by the series of measuring observations and calculations of which they are the result." (71) No place here for transcendence, that "more" which is the hallmark of reality. The thing has been sharply defined, cut down to size, so that it can be manipulated with mathematical precision. Those who think of it as "some entity disporting itself in some metaphysical realm of existence," Eddington goes on to say, "do so at their own risk; physics can accept no responsibility for this embellishment." It happens, of course, that physicists do tend thus to "embellish" their objects, but this is another matter. What concerns us presently is not the *scientistic* universe—the world as depicted in our contemporary scientific imagination—but the *physical* universe, conceived according to the canons of physics. And in that universe an object *is* what it is defined to be, and nothing more. No room for transcendence here, as I have said; the very nature of the scientific process guarantees as much: this is the price one has to pay for the attainment of mathematical exactitude.

It has long been debated whether mathematical entities exist in some Platonic world of their own; it is however quite evident that they do *not* exist here on earth, that is to say, in the objective universe. The case is similar, it now turns out, for *physical* objects: these too do not exist objectively, as Eddington discovered by way of his epistemological reflections. This conclusion could have been foreseen: the conspicuous lack of transcendence on the part of physical objects constitutes indeed a sign of subjectivity. Nothing that is "closed," nothing that is strictly finite or rigorously circumscribed, can exist objectively. Transcendence is truly the hallmark of reality, as we have said. Once again: "Exactness is a fake."

The question arises now: If physical things do not exist objectively, in what sense do they exist at all? Must we suppose, perhaps, that they exist simply "in the mind of the physicist"? Could it be Eddington's *second* table (the physical one), in fact, that resides in the limbo of consciousness? Certainly not! As "partly subjective" entities, physical objects must occupy a middle ground, and that middle ground must in fact be specified through the economy of measurement. Lord Kelvin was right: physics *is* the science of measurement. Only it must be understood that measurement is not in truth the discovery of something that already exists—as one had imagined in Lord Kelvin's day—but constitutes a creative act: to measure is to determine, to limit, to impose a bound. It is to be noted, moreover, that the act of measuring is always accomplished by means of a corporeal instrument, and that the outcome or result is always registered in a perceptible state of the measuring instrument (for example, in the position of a pointer on a dial).

What, then, is the physical? Thus posed, the question can be readily answered: the physical, one now sees, is precisely *the measurable*. What else could it be, seeing that it is neither the measuring instrument nor the result of measurement! But what *is* the measurable? Could it be a particle, for instance? It could not be a "classical" particle, certainly, because such a particle is fully determined and consequently *not* measurable: one must remember that measurement entails the *imposition of a bound*. The measurable must therefore be indeterminate, yet capable of receiving determination. Not "fully indeterminate," to be sure; the latter notion would be as chimerical as that of a classical particle, which is its logical opposite. The measurable, thus, must be situated between these two extremes. Though reminiscent in certain respects of Aristotle's *potentiae*, it is more than that; for the measurable in the sense of physics carries a certain tendency, which in fact can be conceived in mathematical terms: as a *probability*, to be exact. The measurable, one finds, is thus in effect a probability. It is by way of probabilities, it turns out, that Nature outwits our two-valued logic of "being" and "nothingness": for indeed, a probability is neither the one nor the other, but does evidently occupy a middle ground.

The notion is philosophically difficult, and even Eddington seems to vacillate in regard to what these probabilities are. At one point, for example, he has this to say: "The introduction of probability into physical theories emphasizes the fact that it is knowledge that is being treated. For probability is an attribute of our knowledge of an event; it does not belong to the event itself, which must certainly occur or not occur." (50) Yet, if probability is indeed an attribute of our knowledge, it should tell us something about the object or system that is known. According to quantum theory, in fact, probabilities are all that we do know in the physical world.[15] As Eddington points out: "Wave mechanics investigates the way in which probability redistributes itself as time elapses; it analyses it into waves and determines the laws of propagation of these waves." (51) One is beginning to think that these waves are real! But almost immediately Eddington adds that "a sudden accession of knowledge—our becoming aware of a new observation—is a discontinuity in the 'world' of probability waves," suggesting once again that these probability waves pertain to our knowledge alone—as if they did not, by that very fact, belong also to the physical world. It needs to be understood, moreover, that the discontinuity in question arises not simply from an "accession of knowledge—our becoming aware of a new observation," but from that new observation itself, whether we become aware of it or not. What counts is the interaction of the measuring instrument with the physical system that is being measured: it is this that causes the so-called collapse of the wave function. Some forty pages later, however, it appears that Eddington adopts at last an unequivocally realist interpretation of probabilities: "I must still keep hammering at the question, What do we really observe?" he writes. "Relativity theory has returned one answer—we only observe *relations*. Quantum theory returns another answer—we only observe *probabilities*." (80) Here we have it: what we "really observe" are probabilities. The circumstance, moreover, that relativity theory gives a different answer does not alter this fact: Eddington clearly accepts quantum theory as the more accurate and indeed the more fundamental of the two theories. The conclusion stands: probabilities are "what we really observe."

But if there are indeed probabilities, there must also be things which are *not* probabilities; a probability is after all the probability of something which itself is not a probability. In addition to probabilities, thus, there are also events "which must certainly occur or not occur," to put it in Eddington's words. A distinction needs therefore to be drawn between two kinds of facts; as Eddington explains:

> Probability is commonly regarded as the antithesis of fact; we say 'This is only a probability, and must not be taken as a fact.' What we mean is that the result of an observation, though undoubtedly a fact in itself, is only valuable scientifically because it informs us of the probability of some other fact. These secondary facts, known to us only through probabilities, form the material to which the generalizations of physics refer. (89)

We must not fail to observe that in this somewhat obscure passage the distinction between the corporeal and the physical domains lies concealed. Eddington alludes to two kinds of facts: the result of an observation, namely, and the so-called "secondary facts, known to us only through probabilities," which are said to "form the material to which the generalizations of physics refer." Now the result of a measurement, as we have seen, is realized in the perceptible state of a corporeal instrument: it is thus a "corporeal" fact, the kind which are not probabilities. The "secondary facts," on the other hand, refer evidently to the physical domain, the domain in which facts *are* probabilities, or are "known to us only through probabilities," which amounts to the same.

The realization that "we only observe probabilities" proves to be basic; and I would point out that Frieden's EPI approach substantiates this most forcefully. The carrier of Fisher information, after all, is none other than a probability distribution. There is an input probability distribution, corresponding to the "bound" information J, and an acquired probability distribution, corresponding to I, from which the measured value is sampled; it is here that probability collapses, so to speak, into fact (fact of the first kind, in Eddington's enumeration). Up to that final point—which is moreover impenetrable to mathematical analysis—the measuring process has to do with probabilities: "The basic elements of EPI theory," writes Frieden, "are probability amplitudes."[16] What takes place in measurement is a transmission of probabilities, which gives rise to a corresponding transmission of Fisher information from J to I; and it is Frieden's monumental discovery that this very process determines the physics of the measured phenomenon in the form of an output law. As Frieden explains: "EPI regards each such equation as describing a kind of 'quantum mechanics' for the particular phenomenon."[17] This holds true, moreover, even when one derives seemingly "non-quantum" results via EPI: the classical electromagnetic four-potential, for instance, can now be regarded as a probability amplitude for photons; or again, the gravitational metric tensor in Einstein's general theory becomes a probability amplitude for so-called gravitons. According to EPI, *all of physics* is a quantum theory; how could it be otherwise if "we only observe probabilities"?

A major question must not be overlooked: If probabilities are what we really observe, they must be at least "partly objective"; but then, if it be true that transcendence is indeed the hallmark of objectivity (as we have said), must they not reveal that partial objectivity in some mark of transcendence? But how can that be, seeing that they are rigorously specified? How can something that is defined mathematically exhibit such a mark? Here again Nature contrives to outwit our simple logic: the probabilities of physics manifest transcendence precisely when they "collapse" into objective fact; it is at this moment that they suddenly and unexpectedly, as it were, reveal their objective side by violating the Schrödinger wave equation, which up to that point they had strictly obeyed.[18] In a "flicker of transcendence," so to speak, they announce their partial objectivity—and yet, at the same time, their partial subjectivity as well: for the statement itself is made, not on the physical plane of probabilities, but on the plane of *corporeal* fact, in a perceptible state of a corporeal instrument. The entire mystery of the physical is in a way revealed in the phenomenon of wave function collapse: no wonder the phenomenon has been pondered and debated for over seventy years! And no wonder, too, that it continues to baffle the physicist: for it happens that the probabilities announce their objectivity precisely by *violating* mathematical equations. In a momentary burst of non-compliance, one could say, they prove to be more than a mathematical abstraction, more than a mere *ens rationis*, more in fact than just an imprint of the "net." By their momentary breach of "exactness," the probabilities of physics attest that they are *not* "a fake."

It behooves us now to return to Eddington's "two tables" to see where the matter finally stands. We have taken the pre-Cartesian and indeed "normal" position, which affirms that the perceived table is situated, not "in consciousness," but in the external world: it is the real table, we contend, the one and only table that objectively exists. But where does this leave the physical table: the one composed of atoms or particles? We have rejected the idea that Eddington's two tables have simply changed places, so that it is now the molecular table instead of the perceived that is relegated to the realm of consciousness. But having rejected that alternative, it now appears that *both* tables are situated in the external world: how can that be? Can we think of the second, perhaps, as a kind of ghost-like double pervading the corporeal imperceptibly? That would be an "embellishment" tantamount to corporealizing the physical. No matter how thin or ethereal we picture the "ghost" to be, it remains perforce a corporeal thing; but "things physical," it turns out, are not corporeal, nor are they, strictly speaking, "things": the fact that they are described in terms of *probabilities* should suffice to make this clear. The idea of the "ghost-like double" must therefore be rejected.

What exists objectively, we have said, is the corporeal table X. The physical table SX, on the other hand, pertains to our *knowledge* of X: our *physical* knowledge, to be exact. As such SX is partly objective, because it refers to X, and partly subjective, because SX is not X itself—is not, in other words, the objectively existent table, the one we can actually write upon or touch. The physicist is looking at X through a "lens" which, on the one hand, blocks out its substantial form and accidental qualities, while it brings into view something concealed in the nature of X, something which pertains to that "hidden immensity" to which I have alluded before. What is it, then, that is brought into view? Is it a property of X? One could say so, provided one bears in mind that the property in question refers to the way X responds under physical interrogation. The physical table SX may thus be conceived as a model enabling us to predict the statistical outcome of measurements performed on X. For example, one might bombard X with a so-called stream of particles and observe its response in the form of an emitted radiation: it is SX, to be sure, that tells us what to expect. There must of course be something in the nature of X that justifies SX as a feasible model, failing which SX would lack objectivity, lack truth; but the fact remains that what exists objectively is not SX, but X. The molecular model may conceivably express all that is "physically relevant" in X; yet it remains but a model, an abstract and highly selective representation. Contrary to what we have been taught in schools and universities, real tables are *not* "made of molecules."

The question remains where to place the probabilities of physics. These probabilities cannot be assigned to a model or representation, because they are "what we really observe." But neither, as we have seen, can they be assigned to the objective universe, which proves to be none other than the corporeal. As Eddington points out, the probabilities of physics are truly facts: but facts "of the second kind." And these facts of the second kind are "partly subjective" inasmuch as they cannot be separated from the economy of measurement: they are facts of the "participatory universe," to use John Wheeler's felicitous phrase.[19] It appears that the enterprise of physics has given rise to an intermediate domain, a new stratum or substratum of reality, if you will. The physicist, it turns out, is not simply an observer, but a creator of secondary realities: he observes by creating, one could almost say. He does not, however, create *ex nihilo*, nor does his creation exist on its own. We may think of it as the response of the corporeal world to questions posed by the experimental physicist. The corporeal world replies—it speaks to us, one might say—but its response is conveyed in the language of probabilities. The "words" spoken by corporeal natures in response to the physicist's intervention *are* probabilities. They are situated, thus, midway between being and nothingness: for indeed, not even Nature can truly create.

Contrary to popular belief, the physical universe does not stand on its own, but constitutes a secondary and indeed a participatory reality, as some of the great physicists have come to realize. What as a rule they fail to recognize, however, is that the primary reality stands literally before our very eyes: that every able-bodied man, woman and child has access to that primary reality by way of the God-given instruments of perception. It is true that we normally make poor use of these instruments, and that our perceptions tend to be superficial and marred. But the fact remains that even so we are perceiving the real: the primary things themselves.[20]

Notes

1. Positrons are counted negatively, so that the total remains unchanged by electron-positron interactions.
2. *Das Naturbild der heutigen Physik* (Hamburg: Rowohlt, 1955), 21.
3. Frieden shows that this "interplay" can actually be conceived as a so-called "zero-sum" game between the physicist seeking information and an opponent reluctant to part with it. As it happens, the opponent representing Nature generally wins.
4. I should add that EPI imposes one further condition: in addition to the stated variational principle it imposes an algebraic condition on I and J. Together the two EPI conditions constitute a "top law" from which most and perhaps all laws of physics can be derived. The EPI principle itself, however, though arrived at by means of heuristic considerations, is not deduced *à la* Eddington through epistemological analysis. Frieden's claims are, in this sense, more modest.
5. This is not to say that there must be some mathematical kinship between Frieden's EPI approach and Eddington's group-theoretic derivations. The only thing these exceedingly different theories have in common—on the surface, at least—is their amazing claim to derive the laws of physics through an analysis of the measuring process.
6. The page numbers in parentheses refer to *The Philosophy of Physical Science* (Cambridge University Press, 1939).
7. A few pages earlier Eddington had this to say regarding the Procrustes tale: "Procrustes, you will remember, stretched or chopped down his guests to fit the bed he had constructed. But perhaps you have not heard the rest of the story. He measured them up before they left the next morning, and wrote a learned paper 'On the Uniformity of Stature of Travellers' for the Anthropological Society of Attica." (110)
8. B. Roy Frieden, *Physics from Fisher Information* (Cambridge University Press, 1998), 276.
9. In this way certain particular groups (such as the group of rotations in six dimensions) enter into the foundations of physics.
10. *The Nature of the Physical World* (London: J. M. Dent, 1935), 5.
11. Ibid., 268.
12. There is little doubt that Eddington was alluding chiefly to G. E. Moore and A. N. Whitehead.
13. See especially James J. Gibson, *The Ecological Approach to Visual Perception* (Hillside, NJ: Erlbaum, 1986), and two insightful articles by E. S. Reed and R. K. Jones: "Gibson's theory of perception: A case of hasty epistemologizing?" *Philosophy of Science*, 1978, 45, 519-530; and "Gibson's ecological revolution

in psychology," *Philosophy of the Social Sciences*, 1979, 9, 1, 89-204.

14. The fact that an object is perceived as transcendent does not of course guarantee its objectivity: I am not denying the possibility of error. There is indeed a signature of transcendence which bespeaks authenticity; but that signature too can be forged well enough to be deceptive. To err is human, after all.
15. Strictly speaking, "all that we do know" are *data values*, as Professor Frieden has kindly pointed out. Probabilities, thus, are not "known," but inferred. They are "all that we *can* infer."
16. Op. cit., 83.
17. Op. cit., 84.
18. According to the EPI approach, the Schrödinger wave equation is not violated in the act of measuring, but "re-initialized," as Professor Frieden points out. Thus, strictly speaking, what is violated is the original Schrödinger trajectory, which suffers a discontinuity.
19. "All things physical," writes physicist John Archibald Wheeler, "are information-theoretic in origin and this is a participatory universe." This remarkable statement is quoted on page 1 of Frieden's book.
20. Goethe knew that very well when he remonstrated against the Newtonian doctrine, but his voice was not heard; it appears that Goethe's strictures against the Newtonian *Weltanschauung* began to be taken seriously only when that worldview itself began to collapse. Since then a small number of physicists—and a much larger number of biologists—have come to appreciate the significance of what Goethe the scientist had to say. Heisenberg, for example, has dealt with this question sympathetically in a lecture entitled "Die Goethesche und die Newtonische Farbenlehre im Lichte der modernen Physik" (see *Wandlungen in den Grundlagen der Naturwissenschaft*, Zürich: Hirzel Verlag, 1949). The work of Henri Bortoft, another physicist and student of David Bohm, is of even greater interest. Following Goethe's approach, Bortoft has discovered how deep human sense perception can reach into the objective world. His book (*The Wholeness of Nature*, Lindisfarne, 1996) provides a valuable introduction to a subject of immense importance.

Chapter IV

Bell's Theorem and the Perennial Ontology

In the present chapter I propose to examine the quantum-mechanical notion of nonlocality in light of the perennial ontology. The text is based upon a lecture, delivered in 1988, in which, perhaps for the first time, an aspect of the quantum-reality problem has been dealt with from a traditional metaphysical point of view.

I will begin by recalling certain well-known quantum facts which came into scientific view near the turn of the century, precisely at the moment when a victorious physics seemed to be closing in upon the long-conjectured atoms and the fundamental particles of which they now proved to be made. What happened in that final moment, so to speak, is that the quarry has mysteriously eluded our grasp. For indeed, in light of experimental scrutiny these so-called fundamental particles exhibited a hitherto unsurmised wave aspect, implying that they were not, strictly speaking, particles in the first place. The underlying reality, therefore, proved to be an unknown, an X, which under appropriate conditions can exhibit both particle and wave characteristics, and hence cannot be conceived in either terms. What this X is in its own right, no one knows; the quarry, as I have said, has quite eluded our grasp.

To be sure, what is actually observed—what registers on our instruments of detection as a condensation track, or a flash of light on a phosphor screen, or a dark spot on a photographic plate—is indeed the particle aspect, or the particle, as we continue to say. The wave aspect, on the other hand, manifests itself indirectly through statistical laws governing the observed behavior of the so-called particles. For instance, if a beam of electrons is passed through a small round aperture, what appears on the phosphor screen, or on the photographic plate, is a pattern of concentric bands (resembling an archery target). Now this is none other than the so-called Airy pattern, named after the British astronomer George Biddel Airy, who demonstrated, back in 1835, that such concentric bands can be explained mathematically in terms of the diffraction and interference of an impinging wave. Thus, in the statistical distribution of the individual electron encounters with the screen or plate, the electron beam acts as if it were a wave. Oddly enough, moreover, one gets exactly the same Airy distribution whether the electrons are passed through the aperture in vast numbers or sparsely, one electron at a time, let us say. A single electron, therefore, has both a wave and a particle aspect. And whereas the particle aspect suggests that the X in question is discrete and localized, the wave aspect, on the other hand, suggests that it must be continuous and somehow spread out over the entire space of the experiment.

Such, in brief, are the quantum facts that have forced an abrupt end to the Newtonian age. It soon became apparent that the classical formalism of mathematical physics was inherently incapable of coping with the phenomenology of the newly-opened quantum world; and in 1925, after a quarter of a century of theoretical chaos, a new and radically different fundamental theory came at last to birth. And this "quantum theory," as it came to be called, has proved to be a brilliant success. It has been applied to countless problems and has never yet failed, never yet given an incorrect answer. It is truly a marvel, a scientific triumph.

However, there is one great problem with this marvelous theory, and that is that by its very form it gives us no information—no inkling, even—about the nature of physical reality as such. The theory tells us with admirable precision what our instruments of detection will register when a beam of electrons, for example, is subjected to certain conditions; but it tells us nothing about the electron as such. We do not know, for instance, whether the electron owns its dynamic attributes (such as position or momentum) *before* it is observed—whether it is thus what physicist term an "ordinary object,"or whether, on the other hand, these dynamic attributes are in fact "contextual" (which would mean that they exist *only* in the context of an actual measurement). It is basically the old conundrum whether external objects exist when there is no one present to perceive them—but with a very modern twist: for indeed, the problem today is whether it is actually possible to conceive of a noumenal quantum reality in such a way as not to contradict what we know about phenomena.

This is the issue that stands at the heart of the so-called quantum-reality problem on which physicists have been sharply divided. Already in

1927, when the heavyweights of physics gathered in Brussels to assess what (if anything) was still left of the Newtonian *Weltanschauung*, Niels Bohr and Albert Einstein took up diametrically opposing positions on this question which have polarized the debate ever since. What Bohr said, in effect, is that when it comes to quantum phenomena, we must be satisfied with the pragmatic kind of information the quantum theory delivers. It is the best one can do, not because the theory as such is imperfect or incomplete (as Einstein charged), but because there is in fact no "deep quantum reality" behind the phenomena which could explain what is going on. "There is no quantum world," said Bohr. "There is only an abstract quantum description."

Einstein, on the other hand, argued with all his considerable might in favor of a realist interpretation. There *is* a "quantum world," he said in effect, a deep reality which underlies the phenomena and ultimately explains them: there must be! It was unthinkable to him that an "abstract quantum description," unsupported by any deep reality, could lead to correct predictions—as if by magic.

As it turned out, most physicists eventually joined the Bohr school of thought, which came to be known as the Copenhagen interpretation of quantum theory. In 1932, moreover, just five years after the Brussels conference, it seemed very much as though the Copenhagenist claims had been vindicated, once and for all, through the labors of a Hungarian mathematician by the name of John von Neumann. What von Neumann had done, first of all, was to put the new quantum theory on a firm mathematical foundation, and then prove, with exemplary rigor, that the idea of a physical reality consisting of "ordinary objects" is mathematically impossible (that is to say, incompatible with the predictions of quantum theory). At this point it appeared to all but a few realist diehards—led by the unconquerable Albert Einstein—that the issue was henceforth closed.

But in fact it was not. The realist position revived, phoenix-like, in 1952, when David Bohm—after extensive conversations with Einstein—succeeded in constructing an objective model of the electron which squared with the exacting demands of quantum theory. In the wake of the celebrated "von Neumann proof," Bohm had apparently accomplished the impossible. It thus became clear that there must be a loophole in von Neumann's argument; and yet the great mathematician had done his work so well that it took another twelve years to find it! The riddle was not resolved until 1964, when a hitherto unknown physicist by the name of John Stewart Bell discovered that von Neumann had unwittingly made one hidden and unwarranted assumption concerning the "ordinary objects" which quantum theory supposedly ruled out. What von Neumann had assumed—and what until then everyone apparently regarded as self-evident—was that the objects in question were "local" entities. Basically this means that they can only communicate with each other via known physical forces: via signals, therefore, which cannot propagate faster than the speed of light. It turns out, moreover, that Bohm was able to get around the von Neumann interdict precisely because his model of the electron did not in fact obey the stipulated locality condition.

But this result, this decisive breakthrough, proved to be only the beginning. Basing himself on the fundamental principles of quantum theory, Bell succeeded in proving that reality—be it ordinary or contextual!—*must* in fact violate von Neumann's locality condition. What Bell proved finally is that the locality condition is not only unwarranted, but actually untenable. In a word, *reality is nonlocal.* This is the amazing discovery which has come to be known as Bell's theorem. It has since been verified experimentally on the basis of an inequality (likewise discovered by Bell) which however does not hinge upon quantum theory; the result, therefore, appears to stand solid as a rock. But whereas most physicists today accept Bell's theorem as well-founded, a few continue to be skeptical and have left no stone unturned in an effort to thwart Bell's conclusion. Such exceptional scrutiny, moreover, is not unwarranted, considering the momentous implications of Bell's claim. Rightly understood, the affirmation of nonlocality revolutionizes our worldview. It is not without reason that Berkeley physicist Henry Stapp refers to Bell's theorem as "the most profound discovery of science."[1]

What makes it so is the fact that its conclusion points beyond the spatio-temporal universe, the domain to which, strictly speaking, physics is confined by the very nature of its *modus operandi*. The astounding fact is that in the form of Bell's theorem physics has declared its own boundedness, its own incapacity to deal with the deeper strata of cosmic reality. Having, for centuries, claimed title to the objective universe in its entirety, physics has now been forced, on its own ground, to relinquish its totalist pretension.

Actually, the Copenhagenist disclaimer amounts to the same; the die, one can see in retrospect, was cast with the discovery of quantum mechanics. What Bell did was simply to isolate and bring into the open a fundamental feature of quantum theory that had engaged physicists from the start: the existence, namely, and indeed the ubiquity, of what are nowadays termed nonlocal connections. This is what Einstein had in effect observed long ago, and what he took to be a fatal flaw of the new physics. That seeming flaw, however, appears now as a major recognition, a profound truth indigenous to the quantum realm. It is moreover by virtue of Einstein's own insight that one is able to grasp the radical implication of the new discovery; for it is precisely in light of Einsteinian relativity that the trans-spatio-temporal nature of nonlocality can be discerned. Einstein's indictment of quantum theory, one can see, was based in part upon his perception of space-time as the full locus of cosmic reality.

That "classical" presumption has now been negated by Bell's theorem; and as I have said before, physics has thereby declared its own boundedness, its own incapacity to deal with the deeper strata of cosmic reality. It appears that physics as such lacks both the intellectual tools and the requisite data for this properly ontological task. Its domain has thus shrunk, as it were, from the universe as such to a mere class of phenomena. As one unusually astute and forthright quantum physicist has put it: "One of the best-kept secrets of science is that physicists have lost their grip on reality."[2]

What is in fact required if one would "grasp reality" are certain conceptions and insights pertaining to the traditional metaphysical wisdom of mankind, which I would like now to place before you. We will take as our starting point the opening verse of Genesis: *In principio creavit Deus caelum et terram.* First of all, it needs to be clearly understood that the "beginning" in which God is said to have created the world is *not* to be conceived in temporal terms. It is definitely not "a moment of time"; for as St. Augustine has observed, "Beyond all doubt the world was not made *in* time, but *with* time."[3] What strikes us next, moreover, is the second half of the biblical verse: what God creates is "heaven and earth"—not one thing, therefore, but as it were, *two*. One can say that the "heaven" and "earth" of the Genesis passage refer respectively to the spiritual and the material poles of creation, or more concretely, to a spiritual and a corporeal domain. The point, in any case, is not that God created two separate worlds—a heavenly and an earthly—but that the cosmos in its integrality comprises two poles or parts. After all, if "heaven" and "earth" were brought into existence in one and the same creative act (as the biblical verse affirms), it stands to reason that despite their ontological disjunction the two must combine so as to constitute a single creation, a single organic whole.

At this point, however, we need to avail ourselves of another fundamental insight: there must also be an intermediary principle or realm—a *metaxy* in the Platonic sense—for as Plato was perhaps the first to observe, "That two things of themselves form a good union is impossible."[4] And so, too, St. Thomas Aquinas declares: "The order of reality is found to be such that it is impossible to reach one end from the other without passing through the middle." One finds, therefore, that the cosmos in its integrality must comprise at least *three* ontological levels or degrees: the spiritual, the corporeal, and the intermediary, as one can say. The creation is thus to be conceived as a hierarchy made up of three principal levels or tiers. The entire structure, moreover, can be most fittingly represented in the form of a symbolic circle, in which the center (or if you will, a central disc) represents the spiritual world, the circumference represents the corporeal or "visible" universe, and the annular region in between represents the *metaxy* or intermediary domain. I would like to add that this symbolic circle—this veritable icon of the *cosmologia perennis*—was known to every major civilization. The great exception, of course, is our own—this profane post-medieval civilization, intellectually dominated by science, which has in effect reduced the cosmos to its lowest tier.

Having recognized the tripartite division of the integral cosmos, we need now to ask ourselves what it is that actually differentiates the intermediary from the corporeal plane. The answer to this question is quite simple, at least in principle: what distinguishes the corporeal domain from the intermediary is the fact that the former is subject to certain quantitative conditions or bounds from which the latter is exempt. These bounds, moreover, constitute the basic determinations which permit us to speak of space, time, and matter (or energy) in a precise quantitative sense. Think of a boundless expanse of water; and then suppose that a pot, having a certain size and shape, is immersed therein. Obviously, the vessel divides what before was undivided, and introduces quantitative determinations where before there were none. We have here a simple model, if you will, a paradigm which enables us to understand, in the first place, that the substantial reality is one and the same on both ontological planes: water, after all, does not cease to be water when it is poured into a pot. Clearly, the "vessel" does not affect the substance; it only creates certain quantitative attributes—accidents in the Aristotelian sense.

Now, it needs to be clearly understood that the primary determinations which characterize the corporeal plane of existence are by no means subjective, or man-made, but derive precisely from the Fiat of God. These primary bounds have been imposed, figuratively speaking, by the Creator and Architect of the universe when He "set his compass upon the face of the deep," as the Book of Proverbs has beautifully expressed it. We need not concern ourselves with the question as to how, based upon the primary or God-given determinations, the *modus operandi* of physics give rise to secondary quantitative determinations, which can then be described in mathematical terms. Suffice it to say that the quantitative determinations which constitute the immediate object of modern physics refer ultimately to the primary cosmic bounds that determine or define the corporeal plane of existence. Thus, in the terms of our ontological metaphor, one can say that physics deals with the container as opposed to the content: it deals with the size and shape of the vessel, but can tell us nothing concerning the nature of the water contained therein.

One must remember that space, time and matter (in the accustomed sense) exist only on the corporeal plane; for indeed, as we have just noted, these are precisely the categories in terms of which corporeal existence as such is defined. On the other hand, there is no doubt a sense in which the notions of space, time and matter apply to the intermediary domain as well; but the point is that these higher modes of space, time and matter are not subject to the customary bounds, and are consequently very hard for us to conceive. By comparison to the corporeal domain, the intermediary world is "unobstructed" in a manner which does indeed defy our ordinary means of comprehension.

We are ready at this point to return to the problematic of quantum theory. "Everything we know about Nature," writes Henry Stapp, "is in accord with the idea that the fundamental process in Nature lies outside space-time but generates events that can be located in space-time."[5] Can it be said, one is now constrained to ask, that "the fundamental process in Nature," which proves not to be subject to the ordinary bounds of space and time, is in fact situated on the intermediary plane? Can it be said, thus, that physics, on its most fundamental level, has in a way rediscovered the intermediary world? One sees that "in a way" it has. Not, to be sure, in the sense of a discovery or theorem that can be formulated on the level of physics itself. What physics *can* prove, and what it has indeed established beyond reasonable doubt, is that external reality, and thus the cosmos as such, cannot be confined within the bounds of Einsteinian space-time: for if it could be thus confined, it would satisfy the Einsteinian condition of locality, which in fact it does not obey. This is all physics as such can tell us; and let us note that in so doing, it has abdicated its former claim to comprehend reality, for it is clear that a reality which is *not* subject to the bounds of space and time lies *ipso facto* beyond the grasp of physical theory. It is true: physicists *have* "lost their grip on reality"; only one should add that in truth they never *had* such a "grip" in the first place. What physics deals with, in the final count, is not the "content," but the "container," to put it in terms of our metaphor.

It is needful, then, to consider the matter at hand from a traditional ontological point of vantage. We have asked whether quantum mechanics has "in a way rediscovered the intermediary world," and to be sure, the very question is meaningless from a strictly scientific point of view. From an ontological standpoint, on the other hand, the question is not only meaningful, but can indeed be answered in the affirmative: to speak of a cosmic reality beyond the bounds of space and time—of a "process in Nature" outside of space-time—is most assuredly to situate that reality, that "process in Nature," on the intermediary plane.[6] What validates the inference is simply the fact that the things of the spatio-temporal world are indeed none other than the things of the intermediary subjected to spatio-temporal bounds; remove these bounds, and what remains is a thing of the intermediary plane. And I might mention in passing that the transition from the former to the latter is implied in the alchemical *solve*, even as the reverse transition is implied in the complementary *coagula*: like the reality itself, the operations of alchemy do not fit into the spatio-temporal stratum of the universe. The upshot, then, is this: Whatever the quantum physicist himself may think, he *has* in fact rediscovered the intermediary world.

The question remains what the physicist does think apropos of nonlocality. Now to be sure, most physicists do not "think" at all: they simply accept the quantum facts and go about their business. One knows very well, moreover, what that business is: its object is to discover new facts of the same generic kind, and then to comprehend all these findings, so far as one can, within a single mathematical structure. The question of "being," of ontology—of *Weltanschauung*, properly so called—does not arise at all; only a small minority of physicists, it appears, have seriously pursued these deeper issues. They are the quantum-reality theoreticians; one should however add that though their number be small, their ranks include most of the great founders, men of the stature of Bohr and Einstein, Heisenberg, Schrödinger and Planck. I have already referred to the fact that Einstein and Bohr assumed opposing positions, and that most quantum reality theorists eventually sided with Bohr. The majority came thus to believe that the pre-quantum worldview which most physicists had tacitly assumed needs now to be abandoned; and so began the search for a new philosophical *Ansatz* that could account for the paradoxical findings of quantum theory. A wide variety of philosophical notions have since been pressed into service, including ideas (such as the famed "many worlds" hypothesis) that are truly bizarre; yet it appears that these labors have so far succeeded only in exacerbating the philosophical confusion. It would seem that just about every conceivable remedy has been tried, with a single exception: not one of these philosopher-scientists has placed himself on traditional metaphysical ground.

What is called for, as I have argued repeatedly, are certain speculative keys which the perennial wisdom alone can supply. And these keys are in fact none other than the principles of traditional cosmology, which I have enumerated summarily in the Introduction to this book. In the preceding three chapters we have been primarily concerned with the first of these principles, which is tantamount to the tenet of non-bifurcation: it was the systematic application of this principle that has led us to the ontological distinction between the physical and the corporeal domains, and has made possible the subsequent reinterpretation of physics as such. In the present chapter, which has to do with the quantum-mechanical finding of nonlocality, we have applied the second cosmological principle, which refers to the hierarchic structure of the integral cosmos. The key to the understanding of nonlocality, it turns out, is to be found in the traditional distinction between the corporeal and the intermediary levels of cosmic manifestation, and thus between the *sthûla* or "gross" and the *sûkshma* or "subtle" modes of being, to put it in Vedantic terms. As we have come to see, the "process in Nature" which gives rise to the observable quantum phenomena is in fact "subtle" in the Vedantic sense, and is thus situated, ontologically speaking, "above" the corporeal level. Strangely enough, quantum physics, when interpreted from a metaphysical point of view, distinguishes thus between *three* ontological planes: in ascending order, these are the physical, the corporeal, and the intermediary. To be precise, quantum mechanics affirms the first ontological discontinuity or hiatus through the phenomenon of state vector collapse, and the second through the phenomenon of nonlocality. No wonder these phenomena constitute the great enigmas of quantum theory! It is understandable that both should have mystified physicists from the start; for indeed, neither can be comprehended from the perspective of physics itself. Standing, so to speak, on the corporeal plane, the physicist is looking "downwards" in the direction of the physical: *away* from the ontological center, the pole of "being" and "essence," towards the outermost periphery of the integral cosmos, the so-called *materia secunda* or "container" of our universe. In this perspective he cannot see the corporeal, much less the intermediary degree, which is situated "above" the corporeal. To arrive at an understanding of what quantum theory itself has disclosed, he needs thus to undergo what is indeed an intellectual conversion or *metanoia*: a revolution of 180 degrees, one can say. Meanwhile the fact remains that quantum mechanics has "in its own way" rediscovered the intermediary domain. Huston Smith once remarked that the modern West is the first society to view the corporeal world as a closed system; that error has now been corrected on the strength of physics itself.

In conclusion, I wish to point out that the ontological recognition at which we have arrived has decisive implications concerning the evolutionist debate. What both the Darwinians and most creationists have failed to grasp is that the corporeal universe in its entirety constitutes no more than the outer shell of the integral cosmos, and that the mystery of origins needs to be resolved, not at the periphery, but precisely at the center of the cosmic circle. So long as one thinks that the origin of a plant or an animal can be conceived as a spatio-temporal event one has entirely missed the point. Briefly stated, all corporeal beings have necessarily a double birth: a pre-temporal birth, first of all, in the divine creative Act, plus a temporal birth that marks their entry into the corporeal domain. Now, what the first birth brings into being is what the Latin Fathers have called the *ratio seminale* and the Greeks the *logos spermatikos*, which is not however a corporeal entity—not actually a seed in the sense of biology—but a spiritual seed, one can say. This seed, metaphorically speaking, is sown in the spiritual center, incubated in the intermediary domain, and brought into manifestation on the corporeal. At a certain moment it breaks through, as it were, into space-time and becomes perceptible; and this event constitutes its second birth, its birth in the ordinary sense.

It is to be noted that this conception of a double birth is in fact biblical. The doctrine is clearly implied by verses 4 and 5 in the second chapter of Genesis; the passage (which, needless to say, is generally misunderstood by contemporary exegetes) may be rendered thus: "These are the generations of the heaven and the earth, when they were created in the day that the Lord God made the heaven and the earth, and every plant of the field before it sprang up in the earth, and every herb of the ground before it grew." Now, the first thing to observe is that the inspired author speaks of "the day that the Lord God made"—not only heaven and earth, but "every plant of the field before it sprang up in the earth, and every herb of the ground before it grew"—which seems indeed to contradict the hexaemeron doctrine of Genesis 1: the idea, namely, that God created, first the heaven and the earth, and then other things successively, in six stages or "days." What has happened is that the perspective has changed abruptly: it has become at one stroke supratemporal, that is to say, metaphysical. We meet the same "esoteric" point of view here and there in both the Old and the New Testament; for example, in that magnificent verse in Ecclesiasticus: *Qui vivit in aeternum creavit omnia simul* ("He that liveth in eternity created all things at once"). That is the key: Everything without exception is brought into being in the single instant of the creative Act, which our Genesis text refers to as "the day that the Lord God made the heaven and the earth, and every plant of the field before it sprang up in the earth." Moreover, to spring up in the earth—to protrude, as it were, above the ground and become visible—this refers quite clearly to the second birth, the corporeal birth in space and time.

This, in brief, is the biblical doctrine concerning biogenesis. Unlike "creationism" of the customary kind, it opens ontological vistas unsurmised by the evolutionist, and enables us to perceive the debate about evolution from a higher point of vantage. One now sees at a glance that Darwinian transformism is but a brutal attempt to explain the mystery of origins on the corporeal plane, where precisely it can *not* be resolved. We know today in light of quantum theory that not even an electron fits into that truly narrow world—what to speak, then, of plants, of animals, and above all, of man.

Notes

1. "Bell's Theorem and World Process," *Il Nuovo Cimento*, 29B (1975), 271.
2. Nick Herbert, *Quantum Reality* (Garden City: Doubleday, 1985), 15.
3. *De Civita Dei*, 11.6.
4. *Timaeus*, 31B.
5. "Are Superluminal Connections Necessary?", *Il Nuovo Cimento*, 40B (1977), 191.
6. One may well ask why that "cosmic reality" could not equally well be situated *above* the intermediary, that is to say, in the spiritual realm. In point of fact, it can! However, the condition of nonlocality forces us to ascend one step only: that is the point.

Ascension of Christ by Giotto. Scrovegni Chapel, Padua, Italy. c. 1315.

Chapter V

Celestial Corporeality

Leiblichkeit ist das Ende der Werke Gottes.
Friedrich Christoph Oetinger

Having reflected at length upon the corporeal plane and touched upon the intermediary, it behooves us now to consider the third and highest cosmic domain, which is the spiritual or celestial world. I propose to view this ontological domain in its lowest aspect or modality, which by analogy may be termed corporeal. I propose thus to reflect upon celestial corporeality. And even here, in this seemingly rarefied realm, we shall discover in the end a very real connection with the findings of contemporary physics.

As is well known, Christianity believes not only in the immortality of the human soul, but also in what it terms the resurrection of the body. Somehow the bodies of the deceased will be "raised," and in the process, transformed and glorified. "Behold," St. Paul declares, "I show you a mystery: We shall not all sleep, but we shall all be changed. In a moment, in the twinkling of an eye, at the last trump: for the trumpet shall sound, and the dead shall be raised incorruptible, and we shall be changed. For this corruption must put on incorruption, and this mortal must put on immortality." (I Cor. 15:51-53)

Whatever may be the outlook of other religions on this question, the Christian does not seek a discarnate posthumous state. "For we that are in this tabernacle do groan, being burdened: not for that we would be unclothed, but clothed upon, that mortality might be swallowed up of life." (II Cor. 5:4)

This miracle—this unthinkable prodigy—the Christian maintains, has already taken place: it has occurred in the resurrection of Christ, "the first-born from the dead." (Col. 1:18). The dogma of bodily resurrection, thus, is clearly central to Christianity: "If there is no resurrection of the dead," St. Paul affirms, "then Christ is not risen . . . And if Christ be not raised, your faith is vain." (I Cor. 15:13 & 17)

This teaching has always been "hard," and offensive to the philosophers. It is easy enough to imagine or to conceive of a discarnate state; but the raising of dead bodies—that is something else. No wonder the learned men of Athens shook their heads in disbelief. "And when they heard of the resurrection of the dead," we are told, "some mocked: and others said, We will hear thee again on this matter. So Paul departed from among them." (Acts 17:32.33) Nor did the skepticism and disdain of the intellectuals abate with the spreading of the faith. One is reminded of the Platonist Celsus, who so eloquently berated Christianity: "a religion befitting an earthworm," he called it. What is new, in our day, is that theologians professing to be Catholic have joined the skeptics, and have felt obliged to expunge the offending dogma by one means or another. In so doing, however, they have contradicted the explicit and indeed *de fide* teaching of the Catholic Church, and have in fact struck at the very heart of Christianity. Once again: "If there is no resurrection of the dead, then Christ is not risen . . . And if Christ be not raised, your faith is vain." For the Christian, the stark spectacle of the "empty tomb" is not negotiable. It cannot be written off as a literary device intended to bring home the idea of immortality to unphilosophic minds. According to Christian belief, the "empty tomb" constitutes, on the contrary, an historical fact testifying to the greatest miracle ever witnessed on earth: the bodily resurrection of Christ. Christians believe, moreover, that a similar resurrection will take place universally "at the end of time," when Christ will come again: "in the clouds of heaven, with power and great glory." (Matt. 24:30)

Now, the pressing problem for theology—and indeed for metaphysics—is to elucidate the dogma of bodily resurrection, to render it intellectually palatable, so to speak. What Christ and His Church have taught concerning the nature of God, or of the human soul, is far less difficult, I believe, far less problematic to the human intellect. The hardest dogmas, it seems, are those that involve the notion of *body*, the idea of "flesh and blood" if you will. Think of the words of Christ when He taught in the synagogue at Capernaum: "Verily, verily, I say unto you, Except ye eat the flesh of the Son of man, and drink his blood, ye have no life in you." (John 6:53) The

Jews, moreover, understood well enough that this was more than a mere figure of speech, a mere metaphor: "Many therefore of his disciples, when they heard this, said, This is an hard saying; who can hear it?" The Catholic Church, meanwhile, has continued to insist that the "flesh" and "blood" in question are to be taken literally—with the understanding, of course, that these terms refer precisely to the risen and glorified body of Christ. It is indeed this glorified or celestial body that is offered to the faithful as food in the sacrament of the Eucharist. But let us not forget that the host, when consecrated, does not disappear; it is not simply replaced by the body of Christ, but is transfigured into that body: it is "transubstantiated." What takes place, thus, upon every Catholic altar, is akin to the Resurrection: it is one and the same Christic miracle, one can say, that presents itself in two modes.

The mystery of celestial corporeality and its attainment—be it in the Resurrection or in the Mass—has of course been speculated upon by theologians since apostolic times. Yet it appears that some sixteen centuries into the debate a Christian layman—a cobbler and peddler of clothing, no less—was able to contribute decisive insights, which perhaps to this day have been insufficiently reflected upon. I am referring of course to Jacob Boehme, whose teachings can be viewed as inaugurating a new school of theological speculation. In the present article I propose, first of all, to give a brief overview of Christian thought as it bears upon the nature of celestial corporeality, from its apostolic beginnings up to the end of the sixteenth century. These historical observations, moreover, are based upon a paper by Julius Hamberger, which appeared in the *Jahrbuch für Deutsche Theologie* in 1862.[1] I might add that Hamberger was a disciple of Franz von Baader, the renowned Catholic exponent of Boehme's thought, and was himself a profound thinker, who deserves to be far better known than has so far been the case. After the aforesaid historical overview, I shall sketch the relevant conceptions of Jacob Boehme, following closely the interpretation of Pierre Deghaye.[2] What I hope to convey is the fact that Boehme's doctrine enables us to view the subject of "*himmlische Leiblichkeit*" in a distinctly new light. Following this, I propose to reflect further upon the nature of celestial corporeality, in an effort to distinguish the celestial as clearly as possible from the kind of corporeality known to us here below. The final section will focus upon "time, eternity, and light," and it is here that a connection with relativistic physics will become manifest.

Hamberger begins his historical essay with a summary consideration of "*himmlische Leiblichkeit*" as conceived in various pre-Christian traditions. Considering, however, the extreme difficulty of penetrating, *par distance* and from the outside, teachings such as that of the ancient Egyptians, or of the Hindus, it may be best to leave out of account the views of the German savant regarding these domains. The case is different when it comes to the teachings of Plato and Aristotle, not only because they stand at the beginning of what may be termed the philosophic mode, but also because both have exercised a profound and indeed decisive influence upon the development of Christian thought. It will therefore be appropriate to begin our recapitulation of Hamberger's essay at that point.

His crucial contention is that the doctrines of Plato and of Aristotle—each in its own way—rigorously exclude what he deems to be the authentic Christian conception of celestial corporeality. "Plato no less than Aristotle," he writes, "despite the fact that they recognized a supreme unity, transcending every opposition, as the source of all being, remained nonetheless confined within a dualism of the ideal and the real, of spirit and matter." A permanent and perfect union of spirit and matter, of soul and body, Hamberger believes, is unthinkable according to either philosophy. One is forced to conclude in either case that the opposition between the real and the ideal, the sensible and the intelligible, is transcended nowhere save in the supra-ontological sphere of the Absolute. An elevation of the material to the spiritual plane is out of the question. St. Paul was right: the Christian tenet *is* "foolishness to the Greeks."

The earliest indications in Judaic tradition bearing upon celestial corporeality are found in the accounts of Enoch and of the prophet Elijah. What we are told in the case of Enoch is exceedingly sparse: when he was three hundred and sixty-five years old, we read in Genesis 5, "Enoch walked with God; and he *was* not; for God took him." Having lived an exceptionally pious and virtuous life, "God took him," and he was seen no more. As Hamberger observes: "The body of this patriarch, one may gather, was delivered from earthly existence by being swallowed up, as it were, by the life of the spirit, or better said, taken up into that life and thus brought to transfiguration." A similar event is recounted in the case of Elijah: one day, as the prophet was walking with Elisha, his son, "it came to pass, as they still went on, and talked, that, behold, there appeared a chariot of fire, and horses of fire, and parted them both asunder; and Elijah went up by a whirlwind into heaven." (II Kings 2:11) The graphic and explicit nature of this account suggests that it may refer to an actual experience of a visual kind; and yet, as Hamberger points out: "No pious Israelite could have doubted that, behind what the bodily senses could perceive, an event has taken place that extends into the realm of the invisible, the transformation, namely, of the material corporeality of the prophet into the supra-material, even if that Israelite was unable to understand clearly the essential difference between the two."

The definitive texts, however, on the question of *himmlische Leiblichkeit*, are evidently those of the New Testament relating to the resurrection of Christ. It is here, in this unique event, that celestial corporeality enters, as it were, into our world: enters upon the stage of history, one might say. What happened on that first Easter Sunday is in a sense historical; speaking of the risen Christ, St. Paul assures us that "he was seen of above five hundred brethren at once." It is easy, however, to misinterpret the biblically attested facts and thereby miss the very point of the Resurrection. One needs to realize that this ostensibly historical event marks nonetheless a radical break with the earthly condition. "From that moment on," Hamberger observes, "the Savior belongs no longer to the earthly realm, but to a higher sphere, from whence he reveals himself to his disciples only in sporadic appearances, as if by visitation." The Savior, one may presume, revealed himself to his yet earth-bound disciples in visible and tangible form, in order to convince them that he has indeed risen from the dead; and yet, when he appeared suddenly behind locked doors or ascended miraculously, he likewise demonstrated that his spirit had now attained total dominion over his corporeal nature. "Thus his corporeality," writes Hamberger, "was now evidently supra-material and no longer subject to the bounds of time and space."

All the same, many among the faithful have conceived of the transfigured corporeality of Christ in basically material terms, suitably "thinned or sublimated," as Hamberger notes. This wide-spread and indeed dominant tendency has moreover been complemented by an opposing trend, epitomized by the rigorous spiritualism enunciated by Origen. As Hamberger explains: "The reason for the spiritualistic course which Origen pursued lay no doubt in his concern to purify the conception of spirit from the contamination of every material admixture, which however seemed to him to necessitate the exclusion of all corporeality, because, with Plato, he regarded the perfect reconciliation of spirit and nature, the complete elevation of the second to the first, as impossible. But inasmuch as this spiritualism proved to be incompatible with the true sense of Scripture, there arose against him determined opponents, who sought with great emphasis to defend the biblical realism, yet could not do justice to the biblical truth."

The foremost among these "determined opponents" was no doubt St. Jerome, who championed what might nowadays be termed the fundamentalist position. True flesh, for him, was only "what consists of blood, veins, bones, nerves and the like; also teeth, stomach, and genitals could not be absent in the celestial state." It may however come as a surprise that even St. Augustine, Christian Platonist though he was, should have held similar views on the subject, "though in a more delicate form," as Hamberger remarks. The reason, however, for this apparent incongruity, is not far to seek: "Without doubt," comments Hamberger, "Augustine was entirely right when he refused to view the transfiguration of the body as its cancellation or disappearance; the true character, however, of the transfigured body, which consists in a perfect harmony and concord of nature with spirit, of the real and the ideal, he failed to recognize. Against the former error he was protected by his faithful adherence to the word of Scripture; the latter recognition, on the other hand, may have eluded him on account of his predilection for Platonic philosophy."

Be that as it may, it does appear that Christian speculation regarding celestial corporeality has tended to vacillate somewhat between two extremes: the fundamentalist and the spiritualistic positions, one might say. But as Hamberger points out: "Besides such erroneous deviations to one side or the other, there have never been lacking in the Church teachers who knew how to hold fast to the biblical sense of celestial corporeality, and employ that conception in the unfoldment of systematic theology." An excellent example would be Tertullian. That this Christian writer was far removed from the spiritualistic camp of Origen is obvious from his famous statement: "Who will deny that God is a body, even though He is a spirit." But the very boldness of this assertion makes it clear that Tertullian is also worlds removed from the fundamentalism of St. Jerome. "Nothing was further from Tertullian's mind," writes Hamberger, "when he ascribed corporeality to God, than to conceive of that corporeality after the fashion of our earthly body." And it is interesting to note that St. Augustine, while finding it hard to penetrate the idiom of Tertullian's language, regarded him nonetheless as basically orthodox. Tertullian "did not wish to depart," writes Hamberger, "from what the Bible itself suggests quite clearly concerning the corporeality of God, and also recognized well enough the truth that all reality must be somehow formed, and that in the absence of some kind of a corporeal nature the spirit could not function as such." For Tertullian spirit and body are correlatives; the one demands the other. "All invisible things," he writes, "have for God their body and their form by which they are visible to Him." As Hamberger observes: "One sees clearly that Tertullian was permeated by the thought of supra-material corporeality." It is also evident, however, that this predilection has never been shared by the great majority of theologians.

Yet one finds that the Christological teachings enunciated centuries later at the Council of Chalcedon are indeed concordant with Tertullian's idea of supra-material embodiment. "When namely it is ascertained at Chalcedon," Hamberger writes, "that one cannot, on the one hand, regard the human nature of Christ as having been absorbed by the divine, but neither can one regard it as existing in opposition to the divine, or even, as it were, side by side, but must instead assume that the human nature is entirely permeated by the divine and received into the same, it follows that something similar must then apply in general to celestial corporeality in relation to the life of the spirit. In consequence of the aforesaid conciliar decision, one is not permitted to think that in this elevated state the flesh has disappeared, as if absorbed by the spirit; but neither may one conceive of the transfigured flesh in material terms." It appears, however, that the teaching of Chalcedon was generally viewed exclusively in its Christological setting, and that its implications for celestial corporeality as such have remained by and large unappreciated. St. John Damascene—to cite just one example—does not hesitate to speak of "deification" with reference to the glorified flesh of Christ, but continues all the same to conceive of the celestial bodies of angels and of saints as material entities.

To be sure, there have also been theologians who did perceive the wider implications of the Chalcedonian position. This holds true above all of John Scotus Eriugena, for whom the idea of celestial corporeality occupies a position of central importance. As Hamberger observes in that regard: "Not only does this very conception appear in its highest purity, but also in such a universal manner as to pervade the entire doctrinal system of this great thinker and render possible the scientific understanding of Christian truth in its main elements." What John Scotus explicates with utmost clarity and emphasis is the supra-materiality of the first creation: "God is immortal," he declares, "and what He alone creates, that is likewise immortal." In keeping with this idea, John Scotus conceives of the original state of man in distinctly celestial terms. He does not hesitate, in fact, to apply the dogmatic formulas of Chalcedon to mankind, and indeed, to the entire universe, in their state of perfection. His doctrine, as it applies to the perfected universe at large, constitutes in effect a Chalcedonian cosmology. John Scotus is able to conceive of a universal transfiguration, because he views the matter of our world, not as something absolutely primordial, but as itself derived from an immaterial principle. "There is nothing in human nature," he declares, "that is not spiritual and intelligible; even the substance of the body is intelligible." Materiality makes its appearance only where spirit or will stand in opposition to the archetypal idea, that is to say, only where "sin" has entered. "It is unthinkable," says John Scotus, "that the body was perishable and material before the cause of death and materiality, namely sin, had appeared." This position, however, has eschatological implications; as Hamberger points out: "The re-elevation to supra-materiality becomes thus possible through the conquest of sin, and the way is thereby opened for a return of the world to God, to the end that God may become all in all."

John Scotus conceives the supra-material state to be supra-spatial and supra-temporal as well. As surely as God Himself is above time and space, he maintains, it is certain that, with the elevation of the presently material world, "time, as the measure of motion, will disappear, and so too space, as the separator of things, shall be no more." Space and time are aberrations, one can say, and each in its own way is a "separator." Both are aberrant, moreover, precisely because, in the celestial order, there *are no* separations of that kind. "There is nothing incredible or irrational," writes John Scotus, "in the supposition that intelligible beings unite, so that they are one, while each retains its own characteristics, but in such a way that the lower is contained in the higher." To help us understand, he gives the example of air permeated by the light of the Sun, which yet retains its own substance. "Likewise, I believe," John Scotus goes on to say, "will the corporeal substance go over into the soul, not to perish, but that, having been elevated to a more excellent condition, it shall be preserved. So too one must suppose that the soul, having been received into the intellect, becomes more beautiful and more similar to God. In the same way I think of the entry into God—not perhaps of all, but certainly of rational substances—in whom they shall reach their goal, and in whom they shall all become one."

This mighty vision of God, man, and universe, however, was not shared by many—not, in any case, where corporeal nature is concerned. In Hamberger's words: "As little as Tertullian was able to gain for his profound spiritual intuitions a wide recognition, so too John Scotus would not be able to convey his sublime worldview to his own and future generations as their common heritage." Whatever the reasons for this resistance or neglect may be, John Scotus' lofty conception of celestial corporeality has apparently had little impact upon the subsequent course of mainstream theological speculation.

The case of Albertus Magnus and his illustrious disciple is of course of special interest. Now, it does appear that Albertus was open to John Scotus' idea of celestial corporeality, although it remains unclear whether he was in any way influenced by the Irish theologian. It seems more likely that Albertus may have derived inspiration from Avicebron, whose writings he studied assiduously, and whose thought was in certain respects kindred to the teachings of John Scotus. As Hamberger explains: "Avicebron namely was sharply opposed to the usual assumption that substance composed of matter and form must always and everywhere be thought of as something material. The concept of the material, he taught, rests upon the notion of quantity, whereas substance need not entail quantity, but may well be conceived without quantity, and hence without materiality, while still composed of matter and form." But whereas this position—which apparently Albertus did not oppose—permits the conception of a supra-material corporeality, Albertus himself seems not to have fully availed himself of this option. According to Hamberger, Albertus Magnus attained to the true concept of *himmlische Leiblichkeit* "only in the case of the God-man, and only with difficulty can his utterances concerning the bodies of the risen saints be reconciled with the idea of an actual transfiguration." When it comes to St. Thomas, moreover, Hamberger thinks that on this issue Aquinas reverted more or less to the position of Jerome and Augustine. In support of this contention he points out, for example, that the blessed, according to the teaching of St. Thomas, "shall appear at their resurrection in the age of their youth, namely, in that period of life which stands midway between growth and decline"—a view which admittedly is worlds removed from the thought of John Scotus. Hamberger maintains, however, that this "insufficient, and indeed erroneous conception of the resurrected body" is to be ascribed, not to any deficiency on the part of St. Thomas himself, but to external constraint: "It was not an inner, but an outer fetter that impeded the flight of his thoughts." In any case, the fact remains that the theological direction championed by Tertullian and John Scotus did not flourish—for whatever reason—within the bounds of the official Roman Catholic Church. The further development of this theological direction is rather to be found near the periphery, so to speak, of the official Church, and after the Reformation, chiefly in Protestant lands. One must look for it, not among theologians of ecclesiastic rank, but in the circles of mystics, physicians, and God-fearing philosophers. It was moreover in the teachings of our pious and lowly cobbler that this movement, if we may call it such, has reached what appears to be its fullest expression, and possibly its consummate form.

German theosophy, to be sure, was not born abruptly with Jacob Boehme. The *philosophus teutonicus* did have predecessors, veritable precursors, the most important being doubtless Paracelsus (1493-1541) and Valentin Weigel (1533-1588). Nonetheless, Boehme's thought has the distinctive character and titanic force of a revelation; and as is typically the case with a legacy of this kind, his teaching proves to be *de facto* inexhaustible.

After centuries of commentary and analysis by some of the brightest intellects of the Western world, the mystery and fascination of this enigmatic figure remains undiminished. Today, in fact, standing at a greater distance, we can all the better perceive the striking power and originality of his thought. Many literary strands, many traditions, have met and mingled in Boehme's system; but in the process they have become transformed. What emerges is a coherent body of ideas bearing the distinctive signature of the Teutonic philosopher. At his hands our perceptions of God, man, and the universe—while remaining rigorously Christian and indeed biblical—have been profoundly altered, and ostensibly enlarged in certain respects. If St. Thomas Aquinas has enriched Christianity by "Christianizing" the wisdom of Aristotle, it can perhaps be said that Jacob Boehme has done likewise with reference to the Hermetic tradition. In his writings a profound and long-forgotten "alchemical" philosophy of nature reveals its contours—but in a new context, a new key. As Pierre Deghaye has expressed it: "Transposing philosophy of nature to the level of a supreme knowledge proper to theology, Boehme made of it a *theosophy*."[3] And he goes on to say: "In fact it is also a theology, that is, a science of God. But it is profoundly different from dogmatic theology, not only Lutheran,[4] but of any confession whatever. Theosophy represents another approach to God . . ."

The genre itself did not originate with Boehme; for as Deghaye points out, Jewish Kabbalah and Islamic mysticism are likewise theosophies. "What these three types of thought have in common," he writes, "is that their subject matter is God making himself known." Boehme, for his part, distinguishes clearly between God as Absolute—what he signifies by the term *Ungrund*—and God as He is revealed, not only to humanity, but first of all, to Himself. There can be no science, Boehme believes, no knowledge of any kind, of the *Ungrund* as such. Only the self-manifesting God can be the object of knowledge, and indeed, of love. Basing himself, one has reason to believe, upon a mystical experience, Boehme envisages a process of self-revelation *in divinis* which, theosophically speaking, "gives birth to God." It is this "eternal birth," moreover, that constitutes the primary theme of Boehme's theosophy. Supra-temporal though it be, Boehme conceives of that "birth" in terms of a sevenfold cycle, which in fact constitutes the archetype of time. As Deghaye observes, it represents "the week of creation transposed to the level of an utterly first origin." Speaking temporally, the cycle can be described as a successive conquest of a primordial darkness, a primordial chaos, which Boehme also calls a dark fire. "Dogmatic theologies," Deghaye points out, "speak first of all of the light that is synonymous with divine perfection. They mention darkness only in reference to the angel precipitated into it. Boehme puts darkness first. The first part of the seven-part cycle of divine manifestation is dark. In order for light to pour forth, it must shatter the darkness." God's birth constitutes thus a Victory, a supreme Heroic Act which prefigures the death and resurrection of Christ. It can indeed be said that "Boehme's God dies before he is born." The central event of Christianity comes thus to be viewed in a strikingly new perspective: "What transpires on earth once Christ has come among us," writes Deghaye, "only objectivizes for us the primordial event that unfolds in the seven-part cycle."

Now, God is revealed—is rendered visible—in His glory, which is a radiant body, a body made of celestial light. We must not however think of the revealed God as existing simply by Himself, "in splendid isolation." As Deghaye explains: "In order to contemplate himself, God requires that a mirror be offered to him in a form he has raised up, which, although inhabited by him, is distinct from him. . . . The mirror is the body of the angels. The revealed God appears the moment angels are there to contemplate him." One sees thus that the birth of God is accompanied by the genesis of His eternal abode, which is the primordial heaven, the celestial realm. Boehme's doctrine is not only a theogony, but a cosmogony as well. It is in fact a triple cosmogony, which is to say that Boehme envisages the genesis of *three* worlds, corresponding to what he terms the three principles. The highest of these is the primordial heaven, which is "a world of light." But this celestial realm, Boehme maintains, is brought into being by way of a sevenfold cycle that begins in darkness: a darkness which, strange to say, already prefigures hell. It is the fall of Lucifer that objectivizes this primordial darkness, and in so doing, gives birth to the infernal realm. It can perhaps be said that Lucifer, in wanting to be "like unto God," has inverted the very process that founds the angelic domain. Instead of light "shattering the darkness," it was now the darkness that shattered the light of his angelic nature; and so "a world of darkness" came to be. Our world, finally, is the third; it is preceded by the creation of Adam and precipitated by his fall. It constitutes, according to theosophy, a temporary world, whose *raison d'être* it is to permit the reconciliation of a fallen humanity with God. By its very nature it constitutes an intermediary realm in which Light and Darkness—the heavenly and the infernal principles—are mingled. Boehme refers to it as the third principle. One might add that this intermediary status, which theosophy ascribes to our world, is indeed borne out by its phenomenology; for as we all know well enough, ours is a world of combat, of ceaseless strife impelled by the clash of opposites.

What is of major concern is the relation of our world to the heavenly, the primordial, which is its archetype. "The first nature," writes Deghaye, "engenders the other nature, our own, which obscures it while at the same time manifesting it sufficiently to reflect it. This second nature will be destroyed [at the end of time] and the primordial nature unveiled, manifesting God in all his Glory." Despite the dissimilarity of the two natures—which in fact puts the first beyond even our imaginative reach[5]—the connection between the two is exceedingly close. As Deghaye has put it: "These two degrees appear successively, but the two natures coexist. Eternal nature remains within the envelope of our nature . . . and yet there cannot be the least confusion between the two." The distinction, however, between "inner kernel" and "outer shell"—the fact that there is not the least confusion between the two—must not be misconstrued as an ordinary separation; one must not forget that "the two natures coexist," and that the second in fact depends upon the first. The "inner kernel," thus, so far from constituting a kind of foreign body, is to be conceived rather as the "kernel of reality," of which the "shell" is but the outer manifestation.

It is becoming apparent that Boehme's doctrine is intimately related to the alchemical quest, and presumably validates whatever truths that age-old tradition may enshrine.[6] One is tempted to surmise, moreover, that Boehme's conception of a primordial and exemplary cycle may indeed harbor a universal science, applicable in principle to every natural domain; and it is not without interest to note that at least one contemporary scientist has made a serious study of Boehme's teaching, and has apparently drawn inspiration therefrom in the pursuit of his own discipline, which is that of particle physics.[7] What Boehme himself has his eye upon, however, is nothing less than the salvation of man, our second birth. This is the one and only "alchemical transformation" the German mystic relentlessly pursues; and he does so as a pious and biblically oriented Christian. "Our second birth," writes Deghaye, "is the equivalent of the resurrection of Christ, and is anticipated in the cycle of primordial nature. What takes place in this exemplary transformation at the threshold of time is already a death and a resurrection. The cycle of origins is simply repeated each time life blossoms forth following transformation. All life is born only to die and be born a second time. The theosophy of Boehme is a theology of the second birth. This is what connects it with Christian mysticism, in which the focal subject is the birth of Christ in us."

Now, all birth is an embodiment, an incarnation, a certain union of spirit with flesh. Not all flesh, however, is of the same kind. St. Paul speaks of this clearly in the fifteenth chapter of First Corinthians, in the celebrated discourse in which he distinguishes between corruptible and incorruptible bodies, the earthly and the heavenly. Our first body is of the earth, earthly; our second shall be celestial, heavenly (*ex ouranou*). If Boehme's doctrine, therefore, is indeed a "theology of the second birth," it is also, by the same token, a theology of celestial corporeality. There is reason to surmise that the concept of *himmlische Leiblichkeit* comes into its own in the theosophy of Jacob Boehme.

To be sure, all bodies—be they ever so celestial!—are made of a substance of some kind. In the case of celestial bodies, moreover, that substance—as one would expect—is the most precious, the most refined, the most excellent of all. "The eternal heaven," writes Deghaye, "of which the bodies of angels are made, is the precious material that develops over the course of a seven-part cycle described as the divine masterpiece." It is best conceived as a pure radiance, a celestial light, of which what we know as light here below is but a pale reflection. This celestial light, moreover, is none other than God's glory, the very radiance that constitutes what theosophy terms the body of God. God's body, thus, as well as the bodies of angels and of the saints in heaven, are composed of one and the same "precious material" born out of primordial nature's dark and fiery womb. As Deghaye says of celestial corporeality: "Light is its substance, its flesh."

We are now, finally, in a position to appreciate the words of Friedrich Christoph Oetinger which I have placed at the head of this article: "Corporeality is the end of the works of God." The consummation of God's works is indeed embodiment: a union of spirit and flesh, in which the two become as one. It is that perfect submission, that complete "transparence" to the indwelling spirit, that is realized in celestial corporeality. As Deghaye profoundly observes: "The only true perfection is that which is incarnated in a body of light."

The first nature," it has been said, "engenders the other nature, our own, which obscures it while at the same time manifesting it sufficiently to reflect it." However, the obscuration of the first nature is such that we are unable—even in our wildest flights of imagination—to picture, or somehow present to ourselves, that radiant and eternal nature which underlies our own. The two natures—the two worlds—actually coexist, which is to say that the first penetrates the second, not spatially, to be sure, but in an ontologic sense. In a word, it lies "within," as Christ Himself has testified: "Neither shall they say, Lo here! or lo there! for, behold, the kingdom of God is within you." (Luke 17:21) That *regnum* or kingdom, moreover, is none other, theosophically speaking, than the first and celestial nature, which lies concealed here below, not only in our human flesh, but indeed in all things. Truly, "The light shineth in darkness, and the darkness comprehended it not." (John 1:5) Boehme may have been the first to penetrate the cosmological sense of this Johannine dictum, which in any case epitomizes his own conception of our universe as the so-called third principle.

Celestial nature, thus, despite its ontological proximity, remains for us literally shrouded in darkness. Yet we incline, almost irresistibly, to picture celestial realities in more or less earthly terms; even eminent theologians, as we have seen, have often enough, apparently, succumbed to this tendency. To do so, however, is to arrive at a conception that is not only false, but in a sense, inverted. As Hamberger has well said:

> It would be totally wrong (*völlig verkehrt*), when conceiving of celestial corporeality, to retain in some way the idea of earthly materiality, in order not to forfeit the nature of corporeality itself. Be it that one imagines that higher corporeality as a perfected earthly substance, elevated to the highest, noblest form, or that one thinks of it as drastically rarefied—neither in the one nor in the other case would one have reached the true conception. The celestial stands simply above the earthly, and every imperfection to which the earthly remains subject, even when sublimated to the highest degree, must be altogether excluded from the celestial. Every earthly admixture, even the faintest breath, that stems from this nether world, would contaminate the thought of celestial nature, and indeed, would straightway cancel the same.[8]

In the terrestrial domain corporeal entities "occupy space," which is to say, admit extension. The question arises, therefore, whether the same applies in the celestial world. Now, it is clear from the start that celestial space—if indeed there be such a thing—does not coincide with the space that we know by way of sense perception. Celestial bodies are not situated, properly speaking, in *our* space; one cannot say, "Lo here! or lo there!" But by the same token, one cannot say of a celestial entity that it is *not* here, or *not* there; for this, too, would be in a way to subject the being in question to the conditions of our space. It is perhaps needless to point out that when one speaks of the celestial realm as residing "within," this too must not be understood in a literal, that is to say, a spatial sense. But whereas celestial bodies thus remain aloof from any and all spatial conditions pertaining to *our* world, it cannot be denied that there must be something in the celestial realm that corresponds to the idea of space and in fact constitutes its archetype.

On this vital question, Julius Hamberger, it seems to me, has something exceedingly important to contribute: "Not infrequently one is reluctant," he writes,

> to attribute extension to celestial formations for fear of degrading them thereby to the level of materiality. Admittedly, an extension such as one finds in earthly things cannot apply to them; however, what is in no sense extended could have no being, no reality at all. Even spirit must have a certain extension; one must not let it be shrunk into a narrow confine, or better said, into the nothingness of a mathematical point, if it is actually to exist. When however earthly formations, on account of the prevailing obstructions affecting their life, move apart into the breadth of terrestrial space, the same cannot apply to the fully vital beings of the celestial realm. Since these admit of no inner separation, their extension must indeed be of an intensive kind, which is to say, they do not extend outwards, but inwards, not in breadth, but in depth; and that depth stands as much above that breadth as eternity looms above time.[9]

The notion of an "intensive extension," of course, is far from clear. And yet it is highly suggestive, so much so that one begins to sense that there *must* be a state answering to a conception of this kind. One is reminded of the "inner space"—the "space within the lotus of the heart"—of which the Chandogya Upanishad speaks, a space said to contain "both heaven and earth, both fire and air, both the sun and the moon, lightning and the stars." In short, all things are to be found in that inner space, a space which, from our mundane point of view, has no extension at all. They are, however, to be found in that interior space, not in their outward manifestation, but in their innermost being, which coincides, as we have seen earlier, with their celestial archetype. The "space within the lotus of the heart" can in truth be none other than celestial space, and if it contains all things, it is by virtue of the fact that all things are indeed celestial at their core.

One sees that the inner space is not like the outer. Given that external space is an image of the celestial, Hamberger's notion of intensive extension suggests that the image is actually inverted. There is not only an analogy but also an opposition between the two. It therefore requires a radical re-orientation—a profound *metanoia* on the part of the percipient—to pass from one kind of spatial perception to the other. And this, to be sure, must be the reason why we find it so difficult—nay, impossible in our present state—to grasp the nature of celestial bodies. As one reads in the Bhagavad Gita: "In that which is night to all beings, the in-gathered man is awake; and where all beings are awake, there is night for the *muni* who sees." No use trying to picture to ourselves what "the in-gathered man"—the *samyamî*—beholds; *his* world is night to us. The fact is that spatial separation or distance as we know it has no place in the celestial realm, in which extension occurs "in depth." One might say: The external space separates, the inner unites; the former is centrifugal, whereas the latter is centered upon God.[10]

An Aristotelian definition will help us to understand this difference more clearly. Quantity, says Aristotle, is that which admits mutually external parts. Take the case of a line segment: not only does a segment, when bisected, break into two mutually external pieces, but these pieces are truly parts. The point is that the whole, in this instance, is indeed the sum of its parts—and not something more, something therefore inherently indivisible. Similar considerations apply to number, that is to say, to discrete as opposed to extended quantity. Even the sophisticated mathematics of our day, in all of its numerous branches, stands yet under the aegis of quantity in the precise Aristotelian sense of the term. And this is only as it should be: mathematics is indeed the science of quantity. Not everything, however, is quantitative. Take color, for example: red or green, obviously, does not admit of mutually external parts. Red bodies, of course, are divisible; but redness is not. One does *not* change the color of an object by breaking it into pieces. Color comes along, if you will, "for the ride," without being in any way affected by division. Color is not a quantity, therefore, but a quality, and thus a thing which cannot be comprehended in mathematical terms. It should be pointed out that the so-called measurement of qualities, their presumed quantification, is not in fact the measurement of a quality, but of a concomitant quantity, precisely. Qualities as such cannot be measured, cannot in any way be quantified.

Now, the point to be made is simply this: The extension of celestial bodies—like the qualities here below—does *not* admit of mutually external parts; it is not divisible. The kind of extension that breaks into separated parts does not exist in the world above; in that realm, extension is intensive, as Hamberger says: it is an extension, not in "breadth," but in "depth." As we pass from our world to the celestial, what is lost is thus precisely the quantitative aspect of things. The qualities, here below, do also, no doubt, bear an earthly stamp; but even so, they derive their essence from the celestial realm. The qualities are like a glass through which the supernal light filters "obscurely" into our world, to use St. Paul's metaphor. The function of quantity, on the other hand, is not in fact to transmit—to convey essence—

but to separate, that is, to externalize. It is by virtue of their quantitative aspect that the things of earth are excluded from the inner world; it is on account of its extension in "breadth" that the camel is unable to pass through the needle's eye. It is also therefore the quantitative aspect of things that will be destroyed at the end of time: "As a vesture shalt Thou fold them up, and they shall be changed." (Heb. 1:12)

The factor that bestows upon the things of this world their distinctive materiality, their earthly cast, is none other than quantity. No wonder we find it hard to let go of quantity—of our quantitative conceptions of space, time, and matter—as indeed we must, if we are to grasp the nature of celestial corporeality. It can be said that all things in this world are rooted in quantity. Like a plant emerging out of the earth, they evolve out of a quantitative substratum, a *materia quantitate signata* that underlies our universe. It is this quantitative substrate, moreover, that becomes objectified through the *modus operandi* of physics and presents itself to the eye of the scientist as the physical universe, his world of atoms and of galaxies.[11] It needs to be clearly understood, however, that this physical universe, which is the intentional object of physics, does not simply coincide with our world, but represents only its quantitative aspect. Physics is blind to the qualities, blind therefore to all essence. It sees only what is indeed outermost, and thus what stands at the furthest remove from the ontologic core. But needless to say, this is not how the physicist generally conceives of his discipline. Trained and conditioned, as he is, to be reductive, to think of the real precisely in quantitative terms, he is prone to identify the physical universe with the world at large. And even when he turns mystical, as he occasionally does, that mysticism itself tends usually to be of a reductive kind. It happens that there is no room in the prevailing scientistic ontology even for such a thing as color,[12] what to speak, then, of celestial realities. Only by way of a radical re-orientation, it appears, can the contemporary scientist—and thus, too, the scientistically conditioned layman—become open once again to the perennial metaphysical wisdom of mankind.

Having reflected at some length upon the nature of celestial space, it now behooves us to address the subject of "celestial time"—which, to be sure, is none other than eternity.[13] It needs to be pointed out, first of all, that even though eternity is in a sense the temporality of the divine life, it differs radically from time, so much so that the notion of duration does not apply.[14] For this reason eternity has often been likened to a moment, which also lacks duration; but whereas the moment of time passes away "in an instant," eternity does not pass at all. It has therefore been termed a *nunc stans*, a "now that stands." All too often, however, it has been conceived as a *static* present, a present in which there can be no movement, and indeed, no life. But such a state, if it could be realized, would be the very antithesis of eternity, which constitutes in truth the fullness and perfection of life: "the perfect possession of an interminable life *all at once*," as Boethius has beautifully said. What the partisans of a static present have failed to realize is that both the past and the future are mysteriously comprehended within the "now" of eternity; the two other "times" are not simply annihilated, but brought into coincidence, so to speak. In eternity "all temporal succession," says Nicholas of Cusa, "coincides in one *eternal now*."[15] Hard as it is for us to grasp the point of such statements, one thing, at least, is clear even so: only thus—only on the basis of such a "coincidence"—can a "perfect possession of an interminable life *all at once*" be conceived. The fact emerges that life alone can be eternal—because eternity *is* life. The life that we know—a life that breaks up into successive periods and terminates in death—is a reduced life, a life no longer whole.[16] True life is by nature eternal, "ever resting in its movement and ever moving in its rest, ever new and yet ever the same," as von Baader says.

We have seen that the substance of celestial nature, its "flesh," is said to be *light*. One may therefore ask what it is that allies light with eternity. There are of course many kinds of light, ranging from the photonic light of the physicist to the "uncreated light" of Orthodox theology, which presumably coincides with celestial light in the sense of Boehme's theosophy. And yet there must be something common to all these modalities, on account of which each can be recognized as an instance of light. Let us therefore consider the light that we know, be it through sense perception or through the *modus operandi* of physics.[17] Now that light, though it belongs yet to the material order, is doubtless the least material of all material entities. As Hamberger points out: "Light stands at the boundary of materiality and can therefore serve as an image of the supra-material, though it is not such itself."

It is astonishing how accurate this observation proves to be when viewed from the standpoint of modern physics. The claim that light stands "at the boundary of materiality" is rigorously confirmed by dynamic as well as geometrical considerations. It follows, first of all, from the fact that photons have zero rest mass, which means that their entire energy is kinetic. One could say that light is a movement, in which there is nothing material—no mass in the ordinary sense—that moves. Light therefore constitutes a limit; it stands indeed "at the boundary of materiality." The same conclusion follows from the fact that the velocity of light is an absolute maximum, unattainable by any other kind of material entity. For the benefit of the mathematically informed, let me note that the 8-dimensional manifold of physically attainable tangents has a boundary made up of light cones. Hamberger's claim proves thus to be correct from a strictly mathematical point of view.

One knows that light is distinguished from all other modes of matter by the mysterious fact that its speed in a vacuum is independent of the reference frame with respect to which it is measured. The so-called speed of light constitutes thus a universal constant, which moreover plays a major role in fundamental physics. What does this indicate? Is not this fact indeed supportive of Boehme's position when he identifies a transcendent mode of light as the primal element underlying the materiality of our world? Given that photonic light, among all the forms of materiality, bears the closest kinship to this primal element, could not the invariance and maximality of its speed reflect the primacy of its archetype? One thing is clear: If photonic light does represent the primal element, this fact cannot but reveal itself on the physical plane in a certain pre-eminence and normative status. One cannot of course predict on metaphysical grounds what specific form that pre-eminence and normative status will assume; but one can indeed, *a posteriori*, recognize the invariance and maximality of light velocity as characteristics of that kind.

It will prove useful to consider what one can say from the standpoint of physics concerning a body "made of light." To be concrete, let us think of this body as a clock, which must then of course be conceived as moving at light velocity. It is true that no such clock can actually be constructed or realized; and yet it can be conceived in terms of a sequence of actual clocks, whose velocities, with respect to some fixed reference frame, approach the speed of light. Now, if we think of a clock as marking time through successive ticks, one can say, on the basis of Lorentz invariance, that these ticks will slow down as the velocity of the clock approaches that of light. In the limiting case, moreover, when the clock has in effect "become light," one is forced to conclude that the ticking has stopped altogether: the "photonic clock" stands still. What does this mean? Can one say that for a body "made of light"—for a photon, in particular—time itself has come to a stop? Can one say that from the moment of its creation to the moment when it is again destroyed, the photon as such does not experience the flow of time? The physicist, of course, can say no such thing. He is concerned, after all, with the results of observations: actual observations, involving actual corporeal instruments. For him the fact that "photonic clocks stand still" can be no more than a formal consequence of Lorentz invariance. And yet, from a metaphysical point of vantage, that formal fact proves to be suggestive in the extreme. The slowing of clocks as light velocity is approached, the fact that at the speed of light "the ticking stops"—I can conceive of no physical scenario that could mirror the metaphysics of eternity more unmistakably. Does not the ticking of a clock signify time? And does not each tick mark a separation of future from past, and represent a present shrunk to a point? What, then, can it mean that a "photonic clock" stands still? It can only mean one thing: For a "body of light" there *is no* separation between future and past, nor is there a separative "now." But clearly, this correspondence between the formal fact and the metaphysics of eternity is too striking to be dismissed. It appears that relativistic physics does testify to a connection between light and the transcendence of time.

But what about the transcendence of space? Should not Lorentz invariance have something to say regarding this question as well? The fact is that it does. Recall the phenomenon of Lorentz contraction: a rod, for example, will contract as its longitudinal velocity is increased. It is to be understood that this contraction is not a compression, caused by some external force, but has to do with the relativity of length, of spatial extension. In the limit, moreover, as the longitudinal velocity of the rod approaches the speed of light, the length of the rod goes to zero. A "photonic rod," thus, has no length, no extension at all. But it is of course an extension of the "separative" kind that is in question here. And thus we find that a "body of light" does exhibit formally the spatial non-separation characteristic of the celestial realm.

There is a further point to be made: Instead of saying that light is in motion relative to non-photonic bodies, it may be more illuminating to say that non-photonic bodies are in motion relative to light. As an image of the supra-material, it is light, after all, that carries normative status. What, then, can it mean that gross material bodies—whatever be their state of motion in relation to each other—are moving at a universal speed of almost exactly 300,000 kilometers per second in relation to light? It seems to me that this curious and exceedingly enigmatic fact can only mean one thing: The universal motion of gross matter relative to light betokens precisely the ontological distance separating gross from celestial corporeality. That basic and universal motion thus constitutes a mark or image, one might say, of what theology terms the Fall, the expulsion from Paradise. That Fall, that Expulsion, is reflected in the very structure of our universe. Not only does photonic light exemplify the eternal, but it appears that a universal speed close to 300,000 kilometers per second bears witness to our expulsion from the celestial state.

"Light," wrote Hamberger, back in 1863, "stands at the boundary of materiality, and can therefore serve as an image of the supra-material, though it is not such itself." Even the physics of light, it turns out a century later, speaks of "the invisible things of God." (Rom. 1:20) Not openly, of course—not as it speaks of physical things—but indeed *en ainigmati*, "as in a riddle." (I Cor. 13:12) And yet physics speaks of celestial things, not vaguely, but in the most accurate manner conceivable. As I have tried to show—using nothing more than Lorentz invariance—photonic light does mirror both the eternity and the intensive extension of celestial corporeality with the utmost precision. To be sure, it is not, strictly speaking, a mathematical precision—since quantity does not apply to the celestial realm—but a precision of the kind appropriate to metaphysical discourse: a precision, namely, based upon analogy. With this understanding, it appears that the physics of light has weighed in on the side of Jacob Boehme and his school on the issue of celestial corporeality.

Notes

1. "Andeutungen zur Geschichte und Kritik des Begriffes der himmlischen Leiblichkeit," Vol. 7, 107-165. I am indebted to Professor Roland Pietsch of the Maximilian University at Munich for bringing this material to my attention.
2. One of the foremost authorities in this field, Deghaye has provided what is perhaps the finest commentary on the thought of the German mystic. See especially *La Naissance de Dieu ou la Doctrine de Jacob Boehme* (Paris: Albin Michel, 1985).
3. Our quotations are from Deghaye's essay "Jacob Boehme and his followers" in *Modern Esoteric Spirituality*, edited by Antoine Faivre and Jacob Needleman (New York: Crossroad, 1995).
4. Though born in Lutheran territory and nominally Lutheran, Boehme was persecuted by the Lutherans, to the point where he could receive Christian burial only through the intervention of an influential friend. But whereas Boehme's doctrine is indeed at odds with the ideas of Luther, it is by no means antagonistic to Catholic theology. Allowing for the fact that theosophy

and dogmatic theology represent different points of view, it can perhaps be argued that Boehme stands within the integral Catholic tradition, which is after all *katholikos*. Admittedly Boehme was no friend of the papacy; but this question has nothing to do with his theosophic insights and need not concern us here.

5. I am leaving out of account what Henri Corbin terms the active imagination, which is an inherently spiritual faculty.

6. Interest in alchemy seems to be nowadays on the rise. Whereas the educated public at large remains no doubt skeptical and indeed disdainful of the ancient discipline, there is today a deepening awareness among the better informed that what stands behind many an "exploded superstition" may be in fact a long-forgotten wisdom. Although Carl Jung was obviously exaggerating when he suggested that four centuries after being expelled from our universities, alchemy stands "knocking at the door," a number of factors have conspired to render the prospect of re-admission less remote, at least, than it had been during the heyday of materialism. In any case, no truly solid grounds for rejecting the ancient doctrine have yet been proposed. Take the case of the so-called four elements: earth, water, air and fire. One can be reasonably certain that these terms were not employed alchemically in their ordinary sense, but were used to designate elements, precisely, out of which substances, as we know them, are constituted. Somewhat like the quarks of modern physics, these elements are not found empirically in isolation, but occur in their multiple combinations, that is to say, as the perceptible substances that constitute what I term the corporeal domain. Now, as I have argued at length in *The Quantum Enigma* (Peru, Illinois: Sherwood Sugden, 1995), corporeal objects are not in fact mere aggregates of quantum particles; and this clearly suggests that there may indeed be elements of the aforesaid kind. It turns out that our habitual opposition to alchemy is based mainly upon scientistic prejudice: upon a reductionist dogma, namely, for which there is in reality no scientific support at all.

7. Basarab Nicolescu, *Science, Meaning and Evolution: The Cosmology of Jacob Boehme* (New York: Parabola Books, 1991). See also my review of Nicolescu's book in *Sophia*, Vol. 3, No. 1 (1997), 172-179.

8. Our quotations from Julius Hamberger in this section are extracted from a second paper by this author dealing with celestial corporeality, also published in *Jahrbuch für Deutsche Theologie*. It is entitled "Die Rationalität des Begriffes der himmlischen Leiblichkeit," and appears in Volume 8 (1863), 433-476. After contributing three major articles on *himmlische Leiblichkeit* to the *Jahrbuch* between 1862 and 1867, Hamberger published this material in book form, under the title *Physica Sacra, oder der Begriff der himmlischen Leiblichkeit und die aus ihm sich ergebenden Aufschlüsse über die Geheimnisse des Christenthums*. The book appeared in Stuttgart in 1869.

9. As another author has observed: "In eternity one thing does not stand outside another, as in our relation of material space, but the one thing is in the other, and yet is different from it" (H. L. Martensen, *Jacob Boehme: His Life and Teaching*, London, 1885, 74). The affinity in all this to the thought of John Scotus Eriugena is very much in evidence.
10. This inversion has been strikingly depicted in Dante's cosmology as portrayed in the *Divine Comedy*. See Chapter 9, especially pp. 173-176.
11. I have dealt with this question in *The Quantum Enigma* (Peru, IL: Sherwood Sugden, 1995), Chapters 1 & 2. For a summary of my position I refer to my paper entitled "From Schrödinger's Cat to Thomistic Ontology," *The Thomist*, Vol. 63 (1999), 49-63, reprinted here as Chapter 2.
12. As I have shown in *The Quantum Enigma*, it is the reductionist interpretation of the so-called secondary qualities (such as color) that renders the findings of quantum theory perplexing and indeed paradoxical. The debate among physicists and philosophers of science concerning "quantum reality"—which began in 1927 and continues to the present day—bears witness to the extent to which the aforesaid reductionism has penetrated the scientific mind.
13. From a Thomistic point of view one would ascribe, not eternity, but "aeviternity" to the celestial realm: a mean, namely, between time and eternity. However, what Boehme terms the birth of celestial nature, and what he terms the birth of God, are conceived to be concomitant and therefore coeval, or better said, co-eternal. One and the same eternity, thus, applies to "the God who reveals Himself" and to His eternal abode. On the other hand, Boehme also speaks of "*Ewigkeit ohne Wesen*," eternity in an absolute sense.
14. As in the case of spatial extension, it is of course possible to transpose the concept of temporal duration to the celestial domain. The point, however, is that the concept, thus transposed, is radically different from that of "duration" in the temporal sense.
15. One is reminded, in this connection, of Mozart's amazing claim to the effect that in the first moment of artistic conception an entire musical composition presents itself unbroken to his inner sense. It is only at a later stage that the work becomes divided into temporally successive parts. These parts, therefore, must have been somehow contained within the original, as perceived by the inner sense, but not as temporally successive, not as "mutually external," but indeed as "coincident."
16. The cause of this impairment, according to theology, is Original Sin. All life here below—including the subhuman—has indeed been maimed.
17. These two forms of light do not actually coincide, but need not be distinguished for the purpose of the present discussion. On this question I refer again to *The Quantum Enigma*, Chapters 1 and 2.

The serpent and the Tree of Knowledge. Codex Vigilanus, Spain.

Chapter VI

The Extrapolated Universe

I propose in the present chapter to reflect upon the discrepancy between contemporary physical cosmology and the Patristic teaching concerning the creation and early history of the world, a doctrine which respects the literal sense of Genesis. One can hardly conceive of two more divergent and indeed antithetical visions of the universe—which of course explains why the Patristic worldview has come to be regarded almost universally as an untenable doctrine, at best a kind of edifying myth. I propose to show, however, that the question is not quite so simple. The problem, it turns out, is intricate, and has rarely if ever been examined in depth.

We need first of all to ask ourselves whether the two divergent visions—the scientific and the Christian—refer indeed to the same cosmos, the same "world"; and surprisingly, perhaps, one finds that in fact they do not. The key recognition proves once again to be the ontological distinction between the physical and the corporeal domains.[1] On the one hand there are corporeal objects: the things, namely, which can be perceived. To obviate a likely misunderstanding, let me state explicitly that these corporeal things belong indeed to the cosmos, the external world, which is to say that from the start I reject the Cartesian premises—reject what Whitehead terms "bifurcation"—and adopt a realist view of sense perception.[2] Having thus regained the *terra firma* of the traditional philosophic schools, one sees immediately that physical science does not in fact refer to the corporeal world; for as we know, it refers ultimately to fundamental particles and their aggregates, things that are categorically imperceptible, and hence not corporeal. These particles and their aggregates constitute thus a second ontological domain, which I designate by the adjective *physical*. It appears that in the course of the twentieth century science has unveiled an imperceptible and hitherto unknown stratum of cosmic reality. Never mind the fact that this remarkable discovery has been almost universally misconstrued on account of a Cartesian bias which in effect denies the corporeal: what concerns us in the present inquiry is that there are these two disparate domains—the physical and the corporeal—and that henceforth every cosmological debate shall need *de jure* to distinguish between these two "worlds." Is it conceivable, then, that the *corporeal* world does in fact accord with the data of Genesis, that is to say, with the Patristic cosmology? I shall argue that this is indeed the case.

The very possibility of a mathematical physics hinges upon the fact that every corporeal object X is associated with a physical object SX from which X formally derives its *quantitative* attributes.[3] One finds, moreover, that X and SX are spatio-temporally coincident, which is to say that the "here and now" of one applies to the other as well—a principle which I have referred to as spatio-temporal continuity.[4] The corporeal domain in its spatio-temporal totality corresponds thus to a certain subset C of the physicist's space-time, a notion which can be made precise.[5] This subset determines moreover a complementary region, which I shall refer to as the *extrapolated* universe. It is mathematically conceivable, of course, that the subset C coincides with space-time at large, which would mean that the complementary set is empty. But it is likewise conceivable that C constitutes but a speck within the space-time of physical cosmology, so that almost the entire universe as contemplated by contemporary science is in fact extrapolated. What, then, are the true spatio-temporal dimensions of C? How far, starting from the verified "here and now," does the corporeal world extend into the billions of years and light years said to span the physical? Now, the first and most basic point that needs to be made in that regard is that physical science as such is unable *in principle* to answer this question. And the reason for this incapacity is quite simple: physical science, by its very nature, is restricted in its purview to the physical domain. What it "sees" through its man-made lenses are physical entities and nothing else; and that is also the reason why, on a fundamental level, the act of measurement—which necessarily involves a transition from the physical to the corporeal—presents itself to the physicist as an enigma, an anomaly bordering upon paradox. The reason why physics is unable to ascertain the dimensions of C and the reason why it cannot understand state vector collapse[6] are one and the same: physics has eyes only for the physical.

What, then, can one do? If physical science cannot enlighten us regarding the spatial and temporal extent of the corporeal world, by what means can this knowledge be gained? By what means, in particular, can we ascertain the *age* of that world? It was St. Thomas Aquinas, let us recall, who raised the question whether reason alone could demonstrate that the world had a temporal beginning; and he found that reason as such is unequal to the task, and that the issue can be resolved only with the aid of Revelation. So too, when it comes to the age of the corporeal world, could it not be Revelation, once again, that holds the key? One knows that Scripture, taken at its word, does in fact answer the question. It informs us that the world is *young*, which is to say that its age is measured—not in millions or billions—but in thousands of years. The exact figure, of course, is hardly at issue, and may indeed be in doubt; there are discrepancies in that regard between the Septuagint and Massoretic texts. But the order of magnitude—which is all that presently concerns us—is *not* in doubt. And so, on this basis, one arrives at the conclusion that the region of space-time corresponding to the corporeal world is indeed a mere speck within the space-time of the big bang model, with its billions of years and light years. Except for that "speck," the universe of contemporary cosmology appears on scriptural grounds to be in fact extrapolated. We shall presently reflect upon the ontological status of the extrapolated cosmos; but first it behooves us to take a closer look at the Patristic doctrine. We need to understand in greater depth what Christianity teaches regarding the nature and history of God's creation.

Before all else it should be pointed out that "free" interpretations of biblical texts are bereft of authority: they amount to little more than private opinions. It is needful to interpret Scripture in light of tradition, or as the Orthodox say, "with the mind of the Fathers," which is indeed, in a mystical sense, "the mind of Christ." Revelation, therefore, resides not simply between the covers of a book, but within the living ambience of the Church, which is truly the Mystical Body of Christ.

My second point is this: Tradition admittedly condones multiple interpretations of biblical texts; however, it is a universal principle of traditional exegesis that the literal or direct sense must be respected. And nowhere, let us add, is this principle more crucial than when it comes to the early chapters of Genesis, where the temptation to discard the literal sense becomes acute. As St. Thomas Aquinas states explicitly in *Summa Theologiae* I.68.1 with reference to the interpretation of Genesis:

> In discussing questions of this kind, two rules are to be observed, as St. Augustine teaches (*Gen. ad lit.*, 1.18). The first is to hold to

> the truth of Scripture (*veritas Scripturae*) without wavering. The second is that since Holy Scripture can be explained in a multiplicity of senses, one should adhere to a particular explanation only in such measure as to be ready to abandon it, if it be proved with certainty to be false . . .

As Etienne Gilson explains: "St. Thomas is here in full agreement with St. Augustine, and expressly claims to have taken from him this double principle," beginning with the precept "to maintain steadfastly the literal truth of Scripture."[7] It appears that Christianity, even in its Scholastic mode, gives preference to the literal or non-metaphoric sense: this is indeed the *veritas Scripturae* which is to be embraced "without wavering." On the other hand, it can also be said that Latin Christianity has at times exhibited a predilection for metaphorical interpretations of biblical texts—not as contradicting the literal sense, but as being "even more true," if one may put it thus. To cite a major example: the interpretation of the Six Days: The Eastern Fathers—from the Cappadocians to St. Chrysostom, St. Symeon the New Theologian, St. Gregory the Sinaite, right up to St. Seraphim of Sarov and St. John of Kronstadt, who reposed in our century—have consistently emphasized the direct interpretation of these "days" as successive periods of short duration, paradigmatic at least of *our* days, in which the stipulated acts of creation took place. Many of the Latin Fathers, on the other hand, beginning with St. Augustine, have opted for a more metaphysical view of creation, according to which the creative act is one and instantaneous, whereas the Six Days refer to successive phases in the temporal manifestation of created beings. One could say that viewed *sub specie aeternitatis*, the six creative deeds of the *hexaemeron* coincide in a single indivisible Act. Whether there be six acts or one depends thus on the point of view: whether one takes one's stand within the cosmos or in "the *nunc stans* of eternity." And I would add that when it comes to cosmology, properly so called, it is the first of these standpoints that takes precedence. It is worth noting too that St. Ambrose (the Father who converted Augustine) stands solidly with the Eastern tradition not only in the way he views the Six Days, but indeed on all major points relating to the interpretation of Genesis.[8] It behooves us now to consider this "cosmological" Patristic doctrine of creation and early history. In the sequel I shall follow the lead of Hieromonk Seraphim Rose, whose pioneering monograph on this subject has proved invaluable.[9]

According to the Fathers, let me reiterate first of all, the Six Days are indeed *days*. The notion that these "days" stand in reality for vast periods of time—a position currently known as the day-age theory, progressive creationism, or old-earth creationism—is altogether foreign to the Patristic mind. It needs however to be pointed out from the start that despite their adherence to the direct or literal meaning of Scripture, the Fathers are by no means "fundamentalists" in the contemporary sense. So far from being simple-minded or naive, I would argue that their doctrine is in fact more subtle—more refined, one could say—than any of the positions commonly espoused in our day. One must not forget that these Fathers were mystics of a high order—saints, no less—and as such were no doubt gifted with certain insights into the state of the world before the Fall. They were able, therefore, to understand—far better than we—the categorical difference between that primeval world and the contemporary universe. The herbs and trees, for instance, said to have flourished in Paradise, are by no means the same as the plants existing in our time; and yet they were truly herbs and trees, the forerunners of herbs and trees existing today. We must remember that with the Fall of Adam the world itself has changed. A drastic corruption has taken place, a change of state which we, endowed with out present faculties—likewise diminished by the Fall—are normally unable to grasp. The Patristic interpretation of Genesis, therefore, literal though it be, is perforce mystical as well. Only when this mystical view of the world prior to the Fall is replaced by a profane understanding does the literal interpretation turn fundamentalist: only then does it become simple-minded indeed and perhaps deserving of the ridicule voiced often enough by men of scientific temperament. Meanwhile however it happens that these sober men of science are guilty of the same offense: they too have imposed the limitations of *their* worldview upon a domain which categorically exceeds these bounds.

According to the Patristic view, the history of the world breaks up into a number of major segments that need to be clearly discerned and carefully distinguished. The first of these periods, to be sure, is given by the Six Days: the days in which the world began. This period is not yet history, properly so called, but the beginning of history; and as Fr. Seraphim points out: "What is the *beginning* of all things but a *miracle*!"[10] By means of suitable hypotheses, contemporary science has succeeded to its own satisfaction in concentrating that miracle at a single point: the so-called big bang. But Genesis informs us that the miracle was spread out, as it were, over a period of Six Days. It is true that "He who liveth in eternity created all things at once" (Ecclus. 18.1); and yet, viewed from the side of the cosmos, that single Act breaks into six successive acts which together make up the *hexaemeron*, the "work of creation." The Six Days, therefore, in which the world began, are profoundly different from all succeeding days, and together constitute a unique and incomparable period of prehistory.

Following upon the Six Days there ensued what could be termed the Age of Paradise, a period which came to an end with the Fall of Adam. And as we have noted before, the Fathers insist that the Genesis account of that period be understood in a sense that is literal and mystical at once. With the Six Days of creation we have still not arrived at *our* world: the observable universe in which we presently find ourselves. To be precise, the Patristic understanding of Paradise is mystical in two respects: first, because it insists that the nature of Paradise exceeds categorically what the "carnal man"—St. Paul's *psychikos anthropos*—is able to comprehend; and secondly, because it claims that the things of Paradise can in fact be "seen" when certain degrees of contemplation have been attained. St. Gregory the Sinaite, for instance, speaks of this explicitly when he explains "the eight primary visions" accompanying the state of perfect prayer, when things previously hidden "are clearly beheld and known by those who have attained by grace complete purity of mind."[11] And I would add that there *are* in fact no other means by which the things of Paradise can be known: the purified mind constitutes the only "telescope," the one and only "scientific instrument," by which these things can be brought within range of human observation.

We learn from Genesis, and from the Fathers, that in the Age of Paradise the state of the world and the conditions of life on Earth were vastly different from what we know today. For example, in that period "the wild beasts of the earth," and "all flying creatures in heaven," and "every reptile creeping on the earth which has in itself the breath of life" (Gen. 1.30) were as yet what we would term "herbivorous." And as to the nature of Adam, the Fathers accept without question that "God created man incorruptible." (Wisdom 2.23) As St. John Chrysostom observes: "Man lived on earth like an Angel; he was in the body, but had no bodily needs."[12] One sees that by contemporary standards the Age of Paradise is still miraculous; only after the Fall—and by virtue of the Fall—do the contours of *our* world begin to come into view, for indeed, "By man came death." (1 Cor. 15.21) Think of it! Where are those Darwinist "chains of descent" entailing death upon death over millions of years! This is a point of capital importance: Whosoever does not accept the dogmatic claim "By man came death" has *ipso facto* rejected the entire Christian worldview. From that point onwards nothing fits anymore: the entire Christian doctrine becomes unraveled. Once that pivotal contention—"By man came death"—has been categorized as an absurdity, the Incarnation, the Redemption, and the Resurrection follow suit. On this point Teilhard de Chardin was not mistaken in the least: Given that the evolutionist account of origins is true, it is needful that the doctrine of Christianity be radically revised: "moved to a new foundation," as Teilhard aptly put it. Basically, one has only two options: to remain with Scripture and the Fathers of the Church, or to cast our lot with the Jesuit paleontologist[13] and his successors; a middle ground, it seems to me, does not exist.

Let us continue: *Our* world—the universe more or less as we know it or normally conceive of it—came into being after the Fall. A cataclysmic transformation of unimaginable proportions took place, which radically altered not only the external world, but man himself, including notably his cognitive faculties. St. Macarius the Great informs us that the bodily expulsion from Paradise had its counterpart in the soul: "That Paradise was closed," he writes, "and that a Cherubim was commanded to prevent men from entering it by a flaming sword: of this we believe that in visible fashion it was just as it is written, and at the same time we find that this occurs mystically in every soul."[14] It was then that the "carnal man," the *psychikos anthropos* "who receiveth not the things of the Spirit of God: for they are foolishness unto him," and who knows them not "because they are spiritually discerned" (1 Cor. 2.14)—it was then that man *as we know him* came into being. Not instantly, in fact, but gradually; for as we also learn from Genesis, Adam and Eve remained for some time "close" to Paradise: close enough to see it from afar. It can further be said, moreover, that mankind has been engaged since primordial times in an ongoing fall from Paradise: every betrayal, large or small—everything that theology knows as "sin"—constitutes a link, as it were, in this (decidedly non-Darwinist!) "chain of descent."[15]

There is reason to believe that even the so-called laws of nature, as we know them, came into force with the Fall; St. Symeon the New Theologian, for example, suggests this quite clearly when he writes: "The words and decrees of God become the laws of nature. Therefore also the decree of God, uttered by Him as a result of the disobedience of the first Adam—that is, the decree to him of death and corruption—became the law of nature, eternal and unalterable."[16] We need however to understand these last two adjectives in a relative sense; for surely St. Symeon understood well enough that these "eternal and unalterable" laws will again be suspended on "the last Day," when "the powers of the heavens will be shaken" (Matt. 24.29), and there will be "new heavens and a new earth; and the former shall be remembered no more . . ." (Is. 65.17) It thus appears that what we know as the laws of nature—what the physicist, for instance, has his eye upon—apply only to an interim phase of the cosmos: to the period, namely, between the Fall and the general Resurrection.[17]

The consequences of this Christian claim are of course incalculable. Inasmuch as the Earth and the heavenly bodies, together with the primary forms of life, including man, came into existence *before* the Fall, the evolutionist scenario—from the big bang to the formation of galaxies, stars and planets, and thence to the formation of increasingly complex molecules, culminating in biological species—this entire scenario, I say, has been cut off at one stroke. To the extent that we have understood "the mind of the Fathers" we realize that the Six Days and the Adamic world before the Fall transcend categorically what "science"—and indeed the *psychikos anthropos*—can ascertain. One does not need to examine in detail the backward extrapolations by which scientists have sought to reconstruct the distant past—right up to the stipulated big bang, and of late, even beyond!—because one knows by way of Scripture and tradition that past a certain point, and well before one exits the integral domain of human history, these extrapolations are bound to be fictitious. One therefore knows from the start that assumptions of a non-verifiable nature must have been smuggled into these extrapolations, and that the observable facts and tested laws of nature do not suffice to yield the evolutionist scenario. We shall return to this question presently.

But first it is needful to observe that Genesis speaks of yet another major turning point in the history of the world, by which the Earth as we know it came to be. I am referring of course to the Flood, which the Genesis account describes in striking detail. We are told, for example, that the waters rose to a height of fifteen cubits—about twenty-two and a half feet—above the highest mountain peak: what are we to make of this? Theologians, to be sure, have tended for about a century and a half to shy away from a literal interpretation of these texts, in fear of rebuke and ridicule from the scientific sector. The Fathers, on the other hand, taking Genesis at its word, have perceived the Flood as a catastrophe of global proportions that has drastically altered the face of the Earth as well as the conditions of terrestrial life. With Noah and his descendants a new phase of history—of human *and* terrestrial history—begins. Even the climatic conditions appear to have radically changed. Before the Flood, one is led to conclude, the distribution of water on Earth was significantly different from what it is today. Apparently a great quantity of moisture was diffused throughout the atmosphere, causing what is nowadays called a greenhouse effect. It was of course in part these "waters above" that came down at the time of the Flood to cause a global inundation. It has been suggested that prior to the Flood direct sunlight could not penetrate to the surface of the Earth, which might account for the fact that the rainbow—the sign of the primordial covenant—appeared for the first time *after* the Flood. One may speculate that during the Flood, and also presumably in the wake of the catastrophe, gigantic upheavals in the tectonic structure of the Earth have taken place, by which the continents and present topography came to exist. We must not forget that the Genesis Flood—if indeed it occurred!—falls in principle within the domain of profane science. Unlike the seemingly miraculous events preceding the Fall, it is said to have taken place upon this fallen Earth some five thousand years ago. Is this a claim, then, which can be reasonably maintained today?

It is to be noted from the start that this is not a matter of merely academic or marginal interest; for in fact the claim is indispensable to the Patristic *Weltanschauung*, which stands or falls as the integral doctrine it is. Here too it can be said: deny a part, and you have denied the whole. I propose now to consider the status of the Patristic doctrine in light of known scientific facts.

Until the early decades of the nineteenth century it was widely believed that the Earth is some seven or eight thousand years old, and that the major fossiliferous strata were deposited at the time of the Flood. By 1900 the accepted age of the Earth had grown to about 100 million years, and today it is five billion, give or take. And to be sure, it is now the generally accepted view that fossiliferous strata were formed over vast periods of geologic history by a more or less uniform process of sedimentation. The shift began in 1830 when Charles Lyell announced the principles of uniformitarian geology, based upon the assumption of an "old Earth." Nine years later John Pye Smith took the next step: declaring the Genesis Flood to have been a local inundation in the region of Mesopotamia, he eliminated in effect the only viable alternative to Lyell's theory. The stage was now set for Charles Darwin, whose magnum opus appeared in 1859.

How, first of all, does one substantiate the old-Earth hypothesis? The principal means of estimating geologic age is by radiometric dating of igneous and metamorphic rocks. The idea is simple: Given that a sample contains traces of a radioactive isotope plus some element belonging to its decay series (commonly referred to as parent and daughter elements, respectively), one can calculate the length of time needed to produce the observed ratio of the two—*provided* one knows the initial ratio, and knows also that no contamination or leaching has taken place during the process of decay. And of course there is a third generic assumption: namely, that the rate of radioactive decay some millions or billions of years ago was the same as it is today. It happens, however, that each of these assumptions is open to serious doubt, especially the first, which seems to be drawn out of thin air precisely to permit radiometric dating. One assumes that initially only the parent element was present; but why should this be the case? One knows, moreover, that usually one or more of these generic assumptions are in fact invalid, for the simple reason that different methods of radiometric dating applied to one and the same sample generally yield significantly different results. In addition, it has been found that radiometric dating, applied to lava samples of known age, can overestimate by incredible factors.

To complicate matters further, it happens that the sedimentary rock which makes up the fossiliferous strata cannot be dated radiometrically at all. It may surprise the non-specialist to learn that these strata are normally dated by means of so-called index fossils, and thus on Darwinist premises! As Edmund M. Spieker—a respected geologist and strict uniformitarian, no less—has observed with reference to the time-scale associated with that famous "geologic column" displayed in every natural history museum:

> And what essentially is this actual time-scale, on what criteria does it rest? When all is winnowed out, and the grain reclaimed from the chaff, it is certain that the grain in the product is mainly the paleontologic record and highly likely that the physical evidence is the chaff.[18]

It appears, moreover, that the hypothetical column was "frozen in essentially its present form by 1840," at a time when geologists had examined only bits of Europe and the eastern fringe of North America. "The followers of the founding fathers," writes Spieker, "went forth across the earth and in Procrustean fashion made it fit the sections they found, even in places where the actual evidence literally proclaimed denial. So flexible and accommodating are the 'facts' of geology." Even so-called disconformities—old strata piled upon young—do not seem to disturb these self-assured geologists: as in the case of Darwinism, so in geochronology it appears that problems can be invariably fixed by adding yet another *ad hoc* hypothesis.

The most embarrassing fact, perhaps, is that the stipulated uniformitarian process of fossil formation seems not to be operating anywhere. As Richard Milton points out:

> Today there are no known *fossiliferous* rocks forming anywhere in the world. There is no shortage of organic remains, no lack of quiet sedimentary marine environments. Indeed there are the bones and shells of millions of creatures available on land and sea. But

> nowhere are these becoming slowly buried in sediments and lithified. They are simply being eroded by wind, tide, weather and predators.[19]

What is evidently needed for fossil formation is *rapid* burial; and as Milton goes on to note: "Not even the most dedicated Darwinist could believe that the average rate of sedimentation of the uniformitarian geologic column (0.2 millimeters per year) is capable of providing rapid burial."

But the worst was yet to come: since 1985 the case for uniformitarian geology has deteriorated dramatically. What has happened is that a French geologist by the name of Guy Berthault initiated controlled experiments designed to ascertain the actual mechanism of sedimentation, which seem to disprove the assumptions upon which Lyell's theory is based. "These experiments," writes Berthault, "contradict the idea of the slow build up of one layer followed by another. The time scale is reduced from hundreds of millions of years to one or more cataclysms producing almost instantaneous laminae."[20] Berthault's results, published between 1986 and 1988 by the French Academy of Sciences, have not unreasonably been referred to as the "death knell" of uniformitarian geology.

Perhaps these few indications, sparse though they be, may suffice to show that geochronology is not the "hard" science one generally takes it to be, and that, when all is said and done, the old-Earth tenet remains today what it has been from the start: an unproved and indeed unprovable hypothesis. An unbiased observer cannot but agree with Milton (himself by no means a young-Earth advocate) when he concludes: "Because radioactive dating methods are scientifically unreliable, it is at present impossible to say with any confidence how old the Earth is."[21] Meanwhile creationists—long disdained and ostracized by the scientific establishment—are conducting respectable geologic research based upon *their* assumptions, which to be sure are inspired, not by dreams of evolution, but by biblical tenets. The hypothesis of a universal Flood as described in Genesis plays of course a central role in these investigations, and appears today to accord far better with the geologic facts than the opposing uniformitarian premise, which can now be safely written off as a discredited surmise. Today one does not need to be a biblical believer to place one's bet on Flood catastrophism.

There can be no doubt that the Genesis account of the cosmos is geocentric. The Earth, we are told in the very first verse, was created "in the beginning," before the Sun, Moon, and stars were made. One is struck by the fact that the heavenly bodies were created on the fourth Day, *after* the Earth had been rendered habitable and plant life had appeared. And when God did create "the lights in the firmament of heaven," He made them "to divide the day from the night," and "for signs and seasons." We are given to understand that the cosmos is indeed Earth-oriented. This pre-eminence of Earth, moreover, mirrors evidently the pre-eminence of man, the theomorphic creature who, as St. Symeon has expressed it, "was placed by the Creator God as an immortal king over the whole of creation."[22] The centrality of the Earth, it appears, is first of all *iconic*: we are not yet in the domain of measurable quantities and strictly geometric relations. As concerns geocentrism in the ordinary sense, it is of interest to note that with the advance of modern civilization the centrality of the Earth has been progressively diminished or obscured, a historical fact which may not be without spiritual significance. As the matter stands today, the Earth has forfeited every trace of its erstwhile pre-eminence: neither in our solar system, nor in our galaxy, much less in the cosmos at large, is its position presently conceived to be special in any way. We must remember, however, that geocentrism refers no doubt to the corporeal as opposed to the physical world; geometrically it has to do thus with the locus of the Earth within the previously defined subset C. And inasmuch as physical science is in principle unable to ascertain the dimensions or boundary of C, it is *ipso facto* incapable of resolving the question of geocentricity as well. Contrary to the prevailing opinion, therefore, geocentrism has not been ruled out of court.

As to the rival concept of heliocentrism, it is debatable whether the Copernican theory as such contradicts the notion of geocentrality. It is a fact that relative to a heliocentric coordinate system—"under this orderly arrangement," as Copernicus writes—"a wonderful symmetry in the universe, and a definite relation of harmony in the motion and magnitude of the orbs, of a kind not possible to obtain in any other way,"[23] is brought to light; but does this prove that it is the Earth and not the Sun that moves? From a mathematical point of view the question does not even make sense: which body moves depends upon our choice of coordinates. Whether we perceive geocentrism to be compromised by the Copernican discovery depends ultimately, not on scientific fact, but on our philosophic orientation. The discovery itself does no more than reveal a hitherto unrecognized "relation of harmony in the motion and magnitude of the orbs"—*ad majorem Dei gloriam,* a Christian can say. No wonder Pope Clement was delighted, and encouraged the Polish savant to publish his findings. The conflict over heliocentrism, as one knows, did not erupt until the following century, when Galileo promulgated, not simply a mathematical theory, but an entire worldview. As I have pointed out repeatedly, it is precisely the bifurcation postulate implicit in Galileo's distinction between the so-called primary and secondary qualities that underlies the scientistic *Weltanschauung* as its basic plank. One sees in retrospect that the famous dispute was not just about astronomical facts, as one likes nowadays to believe, but that underneath the surface a much larger issue was at stake, an issue which vitally affects the Church.

As concerns the contemporary cosmologies, the crucial point to be noted is that these are based not simply on empirical facts and known physical laws, but require in addition a third ingredient of a very different kind: a cosmological model, namely. But whence are these models derived? How, in particular, does one arrive at the model that underlies the most famous cosmology of our day, the so-called big bang cosmology? Here is what Stephen Hawking and George Ellis have to say on this question:

> We are not able to make cosmological models without some admixture of ideology. In the earliest cosmologies, man placed himself in a commanding position at the centre of the universe. Since the time of Copernicus we have been steadily demoted to a medium sized planet going around a medium sized star on the outer edge of a fairly average galaxy, which is itself simply one of a local group of galaxies. Indeed we are now so democratic that we would not claim our position in space is specially distinguished in any way. We shall, following Bondi, call this assumption the *Copernican principle.* A reasonable interpretation of this somewhat vague principle is to understand it as implying that, when viewed on a suitable scale, the universe is approximately spatially homogeneous.[24]

It appears that the geocentric worldview was cast out, not on the strength of scientific facts or known laws of physics, but indeed on *ideological* grounds. The problem with geocentrism is that it smacks of intelligent design. It happens that the founders of modern cosmology espoused an ideology that favors chance or randomness over the notion of an intelligent creator; fundamentally, big bang cosmology is Darwinism on a cosmic scale.[25]

The question arises now whether one can construct a viable relativistic cosmology starting with a *geocentric* model. And a second question is this: In such a geocentric universe, would the Earth be young or old? To be sure, the prospect of a tenable young-Earth cosmology seems rather slim.

Astronomers, we are told, have observed galaxies 12 billion light years distant from our planet: according to biblical chronology, this would mean that the photons responsible for these observations embarked upon their cosmic journey almost 12 billion years *before* the world began! On the face of it, the scenario seems absurd. One must remember, however, that time is not the absolute we normally take it to be, but can run, relativistically speaking, at vastly different rates. Ideally, therefore, it is indeed conceivable that 12 billion years of star-time translates into some thousands of years as measured by clocks on Earth. Given that biblical chronology refers to Earth-time, there would be no inherent contradiction between the postulates of a young creation and ancient stars. But the question remains, of course, whether this conceptual possibility can be realized in a viable relativistic cosmology.

I would like at this point to refer to the remarkable investigations of Russell Humphreys, a physicist with biblical as opposed to Darwinist persuasions, who recently proposed a relativistic cosmology which may prove to be of major interest.[26] Humphreys begins his construction by postulating a bounded spherical space with the Earth at its center. The model is classical: Ptolemaic, almost. But it is also profoundly Christian: the very geometry of such a universe is indicative of design, of purpose—even as the opposing model of an unbounded universe with an "approximately spatially homogeneous" mass distribution is suggestive of randomness, of blind chance. Boundedness itself, moreover, is suggestive of transcendence, a notion closely allied to that of intelligent design, and equally unpalatable to the scientific mainstream. Humphreys nevertheless concurs with the big bang cosmologists on the question of cosmic expansion: this tenet, he believes, is based, not on the so-called Copernican principle, but indeed on empirical grounds. His task then is to construct an expanding spherical cosmos satisfying the equations of general relativity, a problem, it turns out, that relates intimately to the theory of black holes. A black hole, it will be recalled, is a region of space in which the gravitational field is so strong as to cause both matter and radiation inside the region to be permanently trapped: nothing, not even light, can escape—hence the term "black hole." Such a region is bounded by an intangible surface, called the event horizon, at which light rays emanating from inside the black hole are turned back. The event horizon, moreover, is semi-permeable, which is to say that matter and radiation can pass through that surface in one direction: in this instance, from outside in. Stephen Hawking, one may recall, has provided an impressive account of an astronaut traveling toward the event horizon of a black hole;[27] what is striking—apart from the fact that the astronaut is embarked upon a voyage of no return—is the extreme gravitational time dilation[28] experienced as the event horizon is approached:

the astronaut's clock slows down (relative to *our* clocks) to the point where the "ticking" comes to a stop as the event horizon is crossed.

Now, it happens that the equations of general relativity allow also a second scenario, which can be briefly characterized as a black hole operating in reverse. One has once again a region of space bounded by a semi-permeable event horizon; but this time matter and radiation can only pass *out* of the region. These strange solutions were discovered mathematically in the seventies and dubbed "white holes," but up till now have proved to be of little interest to astronomers and cosmologists. It turns out, however, that a white hole is precisely what is needed to obtain a cosmology based upon a spatially bounded model. Humphreys concludes, in fact, that a *bounded* universe *must have* expanded out of a white hole. In the history of a geocentric universe, therefore, there must have been a moment when the event horizon passed through the surface of the Earth; what can we conclude from this? In place of an astronaut approaching the event horizon from the outside, imagine a human observer on Earth emerging from inside the white hole. Now it is Earth-time that almost stands still, while the clocks on distant stars seem by comparison to be racing at a fantastic rate: fast enough, in fact, to measure out a billion years in a period of short duration as registered by *our* clocks. Remarkably, it appears that a geocentric cosmology is necessarily in essence a young-Earth cosmology as well.

It is surely of enormous interest that a geocentric cosmology of the aforesaid kind has been proposed. Creationists are hopeful that white hole cosmology will accord with the observable facts better than the big bang theory, an expectation which might very well prove true. Physical cosmologies, however, no matter how perfectly they may fit the empirical data, are subject to a generic limitation arising from the ontological discrepancy between the physical and the corporeal domains: this is what needs now to be clarified.

First of all, I contend, once again, that the corporeal has ontological primacy over the physical. What does this mean? Obviously, physical entities exist as intentional objects of physics; and as such they have an intentional existence. But so do the objects seen in a dream, or in a mirage, and so, for that matter, do the holes in Swiss cheese. I am not suggesting, to be sure, that physical entities are subjective imaginings, or that they are illusory, or perhaps kindred somehow to holes in cheese! My point, rather, is that physical entities are *derived* in relation to the corporeal. Primary existence, I maintain, stems directly from the creative act of God, and thus from what St. Thomas Aquinas terms *esse*, the act-of-being itself. As we read in the prologue of St. John's Gospel: "All things were made by Him, and without Him was not anything made that was made." Now, cosmologically speaking, "what was made" (on the lowest ontologic level) are corporeal natures, corporeal things; all the rest—from a mirage and holes in cheese to the entities of physics—are derived from these.

This position is of course the exact opposite of the reductionist view, which in effect regards the corporeal as an epiphenomenon of the physical, thus taking the physical as the primary reality. It is however worth noting that the claim of physical primacy is itself opposed to the canons of scientific empiricism, and thus to the dominant trend in the philosophy of science. For indeed, scientific empiricism is epistemological and anti-metaphysical in its orientation: it insists upon the primacy of scientific observation, of controlled sense experience in fact. The claim of physical primacy stands thus in contradiction to the philosophic foundations upon which physics itself is supposedly based. On the other hand, when Niels Bohr declared "There is no quantum world," he was speaking in accord with what has long been the official philosophy of science.

I would also note in passing that one can hardly affirm physical primacy without reifying the physical by endowing it with corporeal attributes. A fundamental particle, thus, comes to be endowed with a shape and position befitting a marble or billiard ball, contrary to what quantum theory permits us to conceive. In a word, in attributing primacy to these putative particles we spuriously corporealize them—a fact which indirectly argues for primacy of the corporeal. As Heisenberg has noted, fundamental particles are not, strictly speaking, "things," but constitute "a strange kind of physical entity midway between possibility and reality."[29] To think of these so-called particles as "things," and indeed, as the primary things, no less—that surely is naive, to say the least.

I contend next that every derived reality derives in fact from a law concerning the primary things. A mirage, for example, derives from a law of optics, and the holes in cheese derive obviously from a law affecting the structure of cheese. So too the physical, I say, derives from a universal law concerning corporeal nature. That law, to be sure, is not a physical law; it is rather the law which founds both physical things and the law or laws which apply to them. Now, it goes without saying that God is the author of corporeal nature and its law; but again, it is necessary to distinguish between corporeal nature before and after the Fall. My point is this: The law that founds the physical domain came into force precisely at the time of the Fall. It can be none other than that "law of nature, eternal and unalterable" of which St. Symeon speaks: the law imposed when God decreed "death and corruption." We must remember—hard as it may be for us to conceive—that prior to this cosmic catastrophe there *was no* death and corruption. It appears that the nature of "flesh," and thus of what we term "matter," has drastically changed. I would point out that we catch a glimpse of this fact in the Gospel accounts of the Resurrection: for example, by the circumstance that the resurrected body of Christ was able to pass through physical barriers such as walls and doors. The entire domain of the authentically miraculous, for that matter, bears witness to the fact that corporeal natures can in principle be freed from the fetters of physical law: that under appropriate conditions these laws can be suspended. And it is of course highly significant that the miraculous abounds within the ambience of mystics and saints: of men and women, that is, who have approached the pristine state, the state of Adam before the Fall.

What, then, does the primacy of the corporeal imply concerning physical cosmology? It forces us to conclude that the physical domain itself came into existence at the time of the Fall, and will cease to exist when "new heavens and a new earth" shall come to be. Thus, as I have suggested before, no physical cosmology retains validity as one extrapolates beyond either of these two God-given bounds of human and cosmic history. Extended beyond either divide, physical theory retains a merely formal sense; in other words, it becomes fictitious. To be sure, such a fiction may yet prove to be of use in organizing or predicting certain facts of observation: the very idea of extrapolation implies as much. One falls into error, however, the moment one regards the extrapolated theory as a factual account of a past or future evolution. No physical theory, in particular, remains factual as one extrapolates backward to the formation of the first man.

A few more words on the subject of white hole cosmology are now in order. The physical theory needs evidently to be "calibrated" by identifying the historical moment at which the shrinking event horizon crosses the surface of the Earth. Humphreys situates this pivotal event at the fourth Day of creation, to coincide with the formation of the Sun, Moon, and stars. He thus extends the physical theory into a zone to which, in light of the foregoing considerations, it does not apply. I would propose instead that the moment in question coincides with the Fall, and thus with the moment at which physical time—time as we know it—begins. That moment, therefore, has no *physical* prehistory. In the language of metaphysics, it marks the passage from aeviternity to time.[30] I find it striking that the physical theory itself bears witness to this fact: for indeed, at that very moment—given the proposed calibration—the gravitational time dilation on Earth reaches its peak. For an instant, as it were, clocks on

Earth stand still: Earth-time comes to a stop. It is the same effect that obtains, according to special relativity, for a hypothetical clock moving at the vacuum speed of light; and as I have argued previously,[31] the phenomenon is indeed reflective of aeviternity.

Physical time—the physical domain itself—comes into being at the moment when the spherical event horizon drops below the surface of the Earth, a moment that coincides with the birth of the event horizon itself. The resultant image of the Earth "bursting through" that surface may indeed be seen as an icon of the Expulsion. Meanwhile time in the firmament is racing at incredible speeds relative to the newly inaugurated Earth-time, and the cosmos itself, centered upon the Earth, is expanding at a fantastic rate. It is as if a gigantic explosion were driving the universe apart. As time goes on, moreover, Earth-clocks also speed up, indicative of the fact that our "distance" from Paradise continues to grow. What I am proposing, thus, is that the rate of clocks can be taken as a measure of that "distance," a distance that is both ontological and geometric. But if that be the case, and if the cosmic expansion is indeed iconic of the Fall, we can readily understand the previously noted connection between geocentrism and young-Earth cosmology: my point is that the Earth, by virtue of its centrality and status as the habitat of man, stands "closer" to Paradise than the outer regions of the cosmos, and that this fact is reflected physically in a corresponding time dilation. It appears that Russell Humphreys has given us a relativistic cosmology which is theologically meaningful and accords with Patristic doctrine.

Yet fascinating as white hole cosmology may be, it too remains inconclusive. To be sure, it is of great interest to know that a geocentric young-Earth cosmology is indeed conceivable from a scientific point of view; but the question remains whether that cosmology is true. Scientifically speaking, what ultimately counts is the empirical evidence; and let us note that a single contrary finding suffices in principle to bring down a theory. Clearly, it is Nature that has the last word. It may therefore be fitting to conclude this brief survey with a little-known tale of scientific discovery bearing upon geochronology, which may in fact prove to be definitive. It would seem that the Creator, perhaps foreknowing the arrogance of man in these latter days, has seen fit to plant bits of evidence in the rocks that found our Earth, which at one stroke disqualify our most prestigious scientific theories.

The story begins in the latter part of the nineteenth century, when geologists commenced to slice up samples of igneous rocks and subject the resultant specimens to microscopic examination. To their surprise they discovered patterns of concentric rings, resembling an archery target, which came to be known initially as "pleochroic halos" on account of their delicate colorations. The origin of these halos remained a mystery until 1907, when John Joly of Trinity College in Dublin demonstrated that they were caused by a grain of radioactive material implanted at their center. It is known that the decay series of uranium 238 contains a number of radioactive isotopes emitting alpha particles: beginning with uranium 238 itself, the series contains uranium 234, thorium, radium, radon, and three isotopes of polonium, all of which are alpha emitters. Each alpha particle will ionize on the order of 100,000 atoms along its path, leaving a permanent trail of dislocations. The length of its trajectory, moreover, depends on its initial energy, which in turn depends upon the source. Since alpha particles are emitted randomly in all directions, each alpha emitter will contribute a spherical shell of corresponding radius to the resultant configuration, which explains the observed pattern. From the sequence of radii one can now ascertain the corresponding portion of the decay sequence. A radio halo (as they are now termed) is thus determined by its first alpha emitter. It is to be noted that the genesis of uranium halos, in particular, poses no great problem. Since uranium has a very long half life (on the order of 4.5 billion years), one can readily conceive how a grain containing this material may become lodged in heated magma, which then cools over thousands of years to form a mineral, which will then, like a photographic plate, register alpha emissions from the remaining uranium, as well as from the radioactive isotopes resulting from its decay. As is to be expected, moreover, these uranium halos retain a radioactive center. The case is different, however, when it comes to polonium halos; for it happens that the half life of polonium is exceedingly short, ranging from 138 days for polonium 210 to 164 microseconds for polonium 214. Any polonium, therefore, implanted in magma, would certainly disappear long before the magma cooled enough to register alpha trajectories. Neither could the polonium have been implanted in the solidified mineral. Nor is it conceivable, finally, that the polonium halos are of so-called secondary origin: that they derive, namely, from a nearby uranium source (as some desperate scientists have proposed). There appears to be no way out: polonium halos—scattered profusely throughout the length and breadth of Precambrian granite—have posed a challenge to evolutionist geology which in the end may prove fatal to the theory. These structures seem in fact to demand nothing less than a young-Earth scenario.[32]

Where, then, does this leave us? What, finally, can we conclude regarding the status of the extrapolated universe as envisioned by mainstream contemporary science? Bearing in mind all that has been said, the answer imposes itself: to put it plainly, that putative universe proves in the end to be metaphysically impossible and scientifically ill-founded to boot.

Notes

1. I have dealt with this question at length in the first three chapters of this book.
2. It is noteworthy that a non-bifurcationist theory of visual perception contradicting the conventional wisdom has recently been proposed by James J. Gibson, a Cornell University professor of psychology, on empirical grounds, no less. See *The Ecological Theory of Visual Perception* (Hillsdale, NJ: Lawrence Erlbaum, 1986).
3. *The Quantum Enigma* (Peru, IL: Sherwood Sugden, 1995), especially 25-27 and 81-83.
4. Ibid., 30-32.
5. The problem is this: If P and Q are points in C, and R is a point "between" P and Q, one wants R to belong to C as well. Now, in a "flat" space-time, one can define C to be simply the smallest convex subset containing all corporeal objects. The general case of a curved space-time can of course be dealt with as well.
6. Ibid., 52-57.
7. *The Christian Philosophy of St. Thomas Aquinas* (University of Notre Dame Press, 1994), 467.
8. His treatise on the subject, entitled *Hexaemeron*, may be found in volume 42 of *The Fathers of the Church* (New York: The Fathers of the Church, Inc., 1961).
9. *Genesis, Creation and Early Man* (Platina, CA: St. Herman of Alaska Brotherhood, 2000). I am profoundly indebted to Fr. Seraphim for opening my eyes to the significance of the Patristic cosmology, and for providing ready access to the relevant source material. For the convenience of the reader I will include page references to Fr. Seraphim's book (using the abbreviation GCEM).

10. GCEM, 397.
11. "Chapters on Commandments and Dogmas" 130, in the Russian *Philokalia*; GCEM, 416.
12. *Homilies on Genesis* 13:4 (in *The Fathers of the Church*, vol. 74); GCEM, 444.
13. For a critical analysis of Teilhard's doctrine I refer to my monograph *Teilhardism and the New Religion* (Rockford, IL: TAN Books, 1988).
14. *Seven Homilies* 4:5 (in Russian); GCEM, 405.
15. I have dealt with this question in *Cosmos and Transcendence* (Peru, IL: Sherwood Sugden, 1984), Chapter 7.
16. *The First-Created Man* (Platina, CA: St. Herman of Alaska Brotherhood, 1994), 82-83.
17. I have argued elsewhere that the basic facts of Lorentz invariance, in particular, are expressive of the Fall: as in an icon, they depict the mark of the Fall upon creation. See Chapter 5.
18. Quoted in J. C. Whitcomb and H. M. Morris, *The Genesis Flood* (Phillipsburg, NJ: P&R Publishing, 1998), 211.
19. *Shattering the Myths of Darwinism* (Rochester, VT: Park Street Press, 1997), 78.
20. Quoted in Richard Milton, ibid., 78.
21. Ibid., xi.
22. *The First-Created Man*, 90; GCEM, 157.
23. *De Revolutionibus* 1:10. Quoted in W. C. Dampier, *A History of Science* (Cambridge University Press, 1948), 111.
24. *The Large Scale Structure of Space-Time* (Cambridge University Press, 1973), 134.
25. We shall have a great deal more to say on these questions in Chapters 7, 8, and 9.
26. *Starlight and Time* (Green Forest, AR: Master Books, 1994).
27. *A Brief History of Time* (London: Bantam Books, 1988), 87-88.
28. Special relativity predicts a "velocity time dilation": the rate at which clocks "tick" diminishes with their velocity and becomes zero at the speed of light. General relativity, which takes account of gravitational fields, predicts also a second kind of time dilation: gravitational fields likewise slow the rate of clocks.
29. *Physics and Philosophy* (New York: Harper & Row, 1962), 41.
30. Aeviternity, properly so called, may be characterized as the temporality appropriate to the celestial or angelic state. But whereas Paradise is situated "below" the celestial state and is already subject to the condition of time, it

yet partakes of aeviternity; as St. Chrysostom observes: "Man lived on earth like an Angel..." And I would note that it is indeed a mean between secular time and aeviternity that is recovered in the liturgical act, and above all in the traditional rite of Mass, in which, as theology teaches, temporal separation is definitively transcended. My position, then, is not that Paradise is aeviternal, but rather that it is reflective of aeviternity. It is in this sense that the moment of the Fall "marks the passage from aeviternity to time": at that moment time ceased to be "liturgical," its link with eternity was broken. And it is by way of this scission, evidently, that "death and corruption" entered the world.

31. Chapter 5, 99-103.

32. A readable account of halo research may be found in Robert V. Gentry, *Creation's Tiny Mystery* (Knoxville: Earth Science Associates, 1992). It is needless to point out that if the shoe had been on the other foot, these findings would have been universally disseminated and held up as "proof positive" of the position they entail.

Chapter VII

The Pitfall of Astrophysical Cosmology

Rumor has it that the big bang is now "a scientifically proven fact." When Arno Penzias and Robert Wilson discovered the since famous microwave background in 1965, the *New York Times* announced the event with the headline: "SIGNALS IMPLY A BIG BANG UNIVERSE." Whether they do or do not is of course another question; but the fact remains that big bang theory has since become the official cosmology. From that time onwards, every college science major has been taught to believe that the universe was born approximately fifteen billion years ago in some kind of an explosion, and has been flying apart ever since. He has been told that this is the reason why stars and galaxies are observed to recede with a velocity proportional to their distance, as the American astronomer Edwin Hubble is said to have shown. And with the aid of ample media coverage, the resultant picture of an expanding "soap-bubble universe" has commended itself in course of time to the public at large. Now, such a radical transformation of our collective *Weltanschauung* cannot but have an impact upon our culture, and above all, upon our religious sensibility—a matter which I propose to examine in due course.

The present chapter breaks into three disparate parts: I will first review the current scientific status of big bang theory, and then proceed to reflect upon the new cosmology from a theological and specifically Christian point of view. I shall contend that despite its seeming affirmation of a *creatio ex nihilo*, that cosmology is in fact profoundly antagonistic to the Christian faith. In the third part I shall reflect upon the claims of contemporary astrophysics as such, without reference to any particular theory or paradigm.

Big bang theoreticians are obviously facing a formidable task; the theory, after all, is obliged to account, at least in suitably rough terms, for the physical evolution of the universe, from what Georges Lemaître termed "the primeval atom" to the vastness of its present state. What is noteworthy, on the other hand, is that one of the most intensive and protracted research endeavors in the history of science has so far succeeded only in exacerbating the difficulties. The story begins, if you will, with Lemaître's "primeval atom" version of big bang theory,[1] presented in 1931 at a scientific conference, and soon rejected by the astrophysics community. Lemaître had anchored his theory to the claim that cosmic rays could only have been produced in the immediate aftermath of the big bang, a conjecture which proved to be mistaken. After a period of inactivity, interest in big bang theory resumed at the end of the Second World War, stimulated conceivably by the spectacle of exploding atom bombs. The second version, in any case, was proposed in 1946 by George Gamov, an unusually charismatic physicist.[2] In place of Lemaître's cosmic rays, Gamov anchored his theory to the chemical elements, which he perceived as a tangible vestige of the big bang. I vividly recall a physics colloquium, at which, to my amazement, Gamov described in detail the nuclear constitution of the universe so many microseconds after the big bang. Nonetheless, his theory also failed: when Fred Hoyle and his collaborators published a paper in 1957, showing that nucleosynthesis in the interior of stars gives rise to heavy nuclei—in proportions comparable to existing values—the second brief era of big bang theory came to an end. For a time it appeared that an alternative cosmology, the so-called steady state theory, had taken the lead. In 1965, however, the tide turned once again—this was the historic moment captured by the *New York Times* in its banner headline.

What, then, is the connection between the microwave background and the big bang? It is clear on the basis of fundamental physics that the big bang event (if it did occur) must have produced an abundance of radiation, which moreover must still exist in the universe (for the simple reason that it has nowhere else to go). Since it must be in a state of thermal equilibrium (again, because there is no "outside" with which it could exchange energy), that radiation is necessarily of a kind emitted by a so-called black body, the temperature of which can be deduced from the frequency distribution. And finally, the radiation field produced by the big bang must be spread evenly throughout the universe (the reason being that homogeneous initial conditions produce homogeneous effects). According to Gamov's original calculations, that radiation field should by now have redshifted to correspond to a black-body temperature of 20°K, which would place the bulk of it in the microwave range. Gamov's temperature estimate was later revised to 30°K by P. J. E. Peebles—and there the matter stood till the day, in 1965, when the microwave background was picked up on a giant "horn antenna" at the Bell Laboratories and deciphered by two young scientists who had never heard of the big bang. Despite the fact that the radiation turned out to have a black-body temperature of 2.7°K (off by a factor of ten), the discovery conveyed the impression that an astounding prediction had now been verified, and that indeed "signals imply a big bang universe."

The theory, however, was not yet in the clear. The biggest problem facing the new cosmology was to account for the large-scale structures of the astronomic universe. And here the microwave background proved to be in fact a formidable obstacle: its very smoothness and isotropy seemed to preclude the kind of "clumpy" universe we now observe. Given that matter in the primeval cosmos was as evenly distributed as the microwave background leads us to believe, how could it become concentrated into stars and galaxies? Presumably some initial fluctuations became amplified under the influence of gravitational forces to form the stellar universe; one finds, however, that the gravitational fields needed to accomplish such a consolidation must be vastly stronger than the total amount of matter in the universe allows. To make things worse, it turns out that relative velocities between nearby stars and galaxies are comparatively small, making it impossible to achieve the larger observed separations within the fifteen or at most twenty billion years allowed by the big bang scenario. Meanwhile the problem has been further exacerbated by a dramatic increase in the dimensions of large-scale stellar objects identified by astronomers. First there were single stars, then galaxies, and then clusters of galaxies; and finally, in 1986, Brent Tully, at the University of Hawaii, discovered that most galaxies within a radius of a billion light-years are concentrated into slender structures measuring about a billion light-years in length. These are the so-called superclusters, which have since been documented by several research teams. In 1990, Margaret Geller and John Huchra of the Harvard Smithsonian observatory discovered a huge band of galaxies, of supercluster magnitude, which came to be known as the Great Wall; and soon thereafter another team discovered a series of similar structures "behind" that so-called Great Wall. Moving outwards (away from Earth), they discovered a sequence of "great walls," more or less evenly spaced about six hundred million light-years apart. This is not at all what big bang theorists expected, or wanted to find. In fact, it is about the worst-case scenario, a discovery to which the *Washington Post* responded with another banner headline: "BIG BANG GOES BUST," this time.

Meanwhile formidable efforts have been expended to examine the microwave background more closely, in the hope of finding anisotropies. One can well understand why cheers rang out when the COBE (Cosmic Background Explorer) satellite disclosed small fluctuations: it was the kind of result the beleaguered theoreticians had been eagerly waiting for. But unfortunately the variations (measuring one part in a hundred thousand) proved to be much too small: it appears that fluctuations on the order of one percent are needed to account for the formation of stellar objects such as Tully's superclusters or the Great Wall. The radiation background is still far too smooth and isotropic to permit an evolution from a postulated "primeval atom" to the observed astrophysical universe.

Big bang theorists typically respond to problems by making additional assumptions. To be sure, a scientific theory does not need to be instantly discarded when it comes in conflict with certain facts of observation: it is normal practice to search for an appropriate hypothesis that can resolve the conflict, a process which often leads to further discoveries.[3] But this fact hardly exonerates a theory which enjoys little or no empirical support, and is kept alive through a proliferation of ad hoc assumptions. I know of only two major scientific theories where such a glaring lack of empirical verification is tolerated: the Darwinian theory of evolution, namely, and big bang cosmology. One might object to my second contention on the grounds that big bang theory did in fact predict the microwave background; not only, however, did it predict a false temperature, but it happens that the microwave background can be cogently explained in other ways.[4] Like Lemaître's "prediction" of cosmic rays, and Gamov's "prediction" of chemical elements, the prediction of a microwave background actually does little to shore up the hypothesis of the big bang. Meanwhile the addition of ever new—and ever more fantastic!—hypotheses to square the original conjecture with the facts is not a good sign; as Brent Tully once put it: "It's disturbing to see that there is a new theory every time there is a new observation."

To illustrate the "logic" of big bang theory, let us consider what one does when it turns out that the amount of matter in the universe is about a hundred times too small to permit the formation of galaxies. Literally with a stroke of the pen one supplies the missing mass—a hundred times the estimated mass in the universe!—by postulating something called "dark matter": a kind which does not interact with electromagnetic fields and has never yet been observed. A profusion of dark matter candidates has been proposed in recent decades by helpful particle physicists: there are axions, higgsinos, photinos, gravitinos, gluinos, preons, pyrgons, maximons, newtorites, quark nuggets, and nuclearites, to mention a few; the problem is that all these exist so far only on paper. But let us suppose that there do actually exist, say, higgsinos or quark nuggets: would this suffice to extricate big bang theory from its quandary? Certainly not; a number of other major problems would remain. What is more, each new hypothesis tends to introduce problems of its own, which will presumably necessitate the introduction of additional hypotheses. It seems unlikely that such a procedure will converge; and if it does, one wonders whether one has *found* a truth, or constructed it, as some postmodernist philosophers of science have claimed.

We have so far considered only one major difficulty: the problem of accounting for the formation of large-scale stellar objects. To round out this brief review, I will recount one more predicament. Let us note, first of all, that when it comes to stellar objects, from stars to galaxies and clusters of galaxies, all that we have to go on is light emitted by the objects in question and received through telescopes, be they terrestrial or mounted on satellites. By light, moreover, I mean electromagnetic radiation of whatever frequency, from radio waves down through the visible range to X-rays and gamma rays. In a word, what we have are light-particles or photons, each of which defines a position on a photographic plate and carries a frequency: that is all. These are, if you will, the actual empirical facts; the rest is theory, a matter of interpretation. The radiation received does however carry a wealth of information, some of which is quite unequivocal in its implications. It is known from laboratory experiments that the distribution of frequencies emitted by a chemical element, for example, is a characteristic of that element. The emission spectrum constitutes thus a signature permitting us to detect the presence of hydrogen, helium, and other elements in stars and galaxies. It happens, however, that spectra received from outer space are generally shifted to the left on the scale of frequencies, a phenomenon known as the redshift; what causes this redshift? It has long been surmised that stellar redshifts constitute a Doppler effect, which is to say that the reduction of frequencies is caused by a recessional velocity of the source; the situation is supposedly analogous to the reduction in the pitch of a train whistle when the train is speeding away from us. On this basis stellar redshifts have been interpreted as evidence for an expanding universe: big bang cosmology is founded upon that hypothesis.

Now, it happens that for the last forty years observational results have accumulated which appear to contradict that assumption. The first bit of bad news for the big bang theorist came in 1963 with the discovery of extragalactic radio sources now known as quasars, whose emission spectrum proved to be heavily redshifted, corresponding to recessional velocities approaching the speed of light. It was soon found, however, that these quasars were typically associated with galaxies whose redshifts are normal, that is to say, small. Stellar objects, which according to big bang geometry were supposed to be billions of light years apart, appeared thus to be close neighbors. The Doppler interpretation of quasar redshifts has therefore become suspect. Meanwhile intrinsic (non-Doppler) redshifts have been surmised in other stellar objects, down to the level of stars; according to one expert, "Most extragalactic objects have intrinsic redshifts."[5] But this would mean that the hypothesis upon which big bang cosmology is based—the assumption that stellar redshift equates to recessional velocity—has been disqualified.

Certain developments on the theoretical side, moreover, have augmented these misgivings. In 1977, Jayant Narlikar, an astrophysicist, succeeded in generalizing the equations of relativity so as to allow the masses of fundamental particles to increase with time. The theory, it turns out, predicts intrinsic redshifts caused by the variation of particle mass. The idea is simple: The smaller the mass of an electron, the smaller will be the energy it loses in a so-called quantum jump, which however is just the amount of energy given off in the emitted photon. Since the frequency of a photon is proportional to its energy, one obtains thus an intrinsic redshift. In place of the redshift-velocity relation underlying big bang cosmology, the Narlikar theory gives us an inverse redshift-age relation, which enables us to interpret existing data in a new key. The Hubble relation, according to which redshift is proportional to distance, can now be understood from the fact that distant stellar objects are observed at an earlier time due to the finite speed of light, and will consequently tend to have smaller particle masses and correspondingly larger redshifts in proportion to their distance. The heavily redshifted quasars, on the other hand, which do not satisfy the Hubble relation, are now perceived as being constituted by recently created particles ejected from an active or so-called Seyfert galaxy. In this way the new theory does justice to all the relevant observational facts: to those which support the Hubble relation, as well as to those which do not. According to the resultant cosmology, moreover, the universe is *not* expanding, and did *not* evolve out of an initial singularity: there *is* no big bang.[6]

It is not my intention to tout the Arp-Narlikar approach, which presumably has problems of its own. My purpose is only to show that one of the major problems facing big bang cosmology is to justify the Doppler interpretation of stellar redshifts, a difficulty which in the end may prove insurmountable.

Many astrophysicists are no doubt disturbed, but few at present are prepared to abandon the official doctrine. Admittedly, a handful of leading scientists have openly proclaimed the demise of big bang cosmology,[7] and even the conservative science journal *Nature* has run a lead editorial under the caption "Down with the Big Bang." But such occasional expressions of dissent have so far had little effect upon the astrophysics establishment; too many careers, it seems, hang in the balance.

Let us now look at the big bang scenario from a theological perspective. Leaving aside the question as to whether this cosmology is factually correct, we shall treat it as a kind of myth or icon, a symbol to be read. What, then, does the big bang signify? What strikes us, above all, is the idea that the universe had a beginning in time: that in a sense the world "did not always exist." This is not to say that there was a time when the world did not exist, for time as we know it refers to cosmic events and cannot therefore antedate the universe itself: "Beyond all doubt," says St. Augustine, "the world was not made *in* time, but *with* time." What big bang theory affirms is simply that the universe has a finite age; the question is whether this implies an act of creation *ex nihilo*. I would argue that, from a strictly logical point of view, it does not. But this is actually beside the point: we are now "reading the icon," a task which is not simply a matter of logical analysis. In its iconic import, I say, the big bang picture does overwhelmingly suggest what Christianity has always taught: namely, that the universe was brought into being some finite time ago through a creative act. As Pope Pius XII declared in 1951, in an address to the Pontifical Academy of Science:

> In fact, it seems that present-day science, with one sweeping step back across millions of centuries, has succeeded in bearing witness to that primordial *Fiat lux* uttered at the moment when, along with matter, there burst forth from nothing a sea of light and radiation . . . Hence, creation took place in time; therefore, there exists a creator; therefore, God exists![8]

It would seem thus that the impact of big bang cosmology upon Christianity is bound to be salutary; but such proves not to be the case. I contend that the new cosmology has exerted a baneful influence upon Christian thought, and has contributed significantly to the deviations and vagaries afflicting contemporary theology; how can this be? The answer is simple: Icons can be dangerous, lethal in fact (it is no wonder, thus, that iconoclastic movements have sprung up time and again). The danger derives from the circumstance that the icon itself can be mistaken for the truth: "the finger for the moon," as the Chinese say. And this is of course what has actually happened in the case of the big bang: we are dealing, after all, with a scientific paradigm declared by the leading authorities to be factually true. Now, the problem is this: In its factual as opposed to its symbolic significance, the big bang scenario is flatly opposed to the traditional Christian cosmogony based upon Genesis. Take for instance the biblical fact that the Earth and its flora were created *before* the Sun, Moon and stars: surely this rules out all contemporary theories of stellar evolution, even as it rules out all Darwinist claims. Theologians, as we know, have for the most part responded to this challenge by "demythologizing" the first three chapters of Genesis; but in so doing, I charge, they have taken a wrong turn. Placing their trust in a man-made theory, which moreover stands demonstrably on shaky ground, they have contradicted the inspired teaching of the Fathers and the Church. I contend that the first three chapters of Genesis, taken in their literal historical sense, cannot be denied without grave injury to the Christian faith. The point has already been made implicitly in the preceding chapter: in bringing to light the content of biblical cosmogony, we have at the same time demonstrated its central importance to Christian doctrine. Whatever contemporary theologians, in their eagerness to become "scientifically correct," may say, the fact remains that the teachings of Christianity presuppose the biblical cosmogony, even as the Redemption presupposes the Fall. It is utterly chimerical, thus, to imagine that the doctrine of Christ makes sense in a big bang universe. And one might add that the biblical cosmogony has in fact been consistently mandated by the Magisterium of the pre-Conciliar Church: for instance, in the response of the Pontifical Biblical Commission to questions regarding "the historical character of the earlier chapters of Genesis." The Commission explicitly denies the validity of "exegetical systems" which exclude the literal historical sense of the first three chapters.[9]

Getting back to big bang cosmology, I would like to point out that this doctrine is evidently all the more compelling to a Christian public on account of its symbolic content: what could be more wonderful, after all, than a scientific cosmology bearing witness to the primordial *Fiat lux*! In conjunction with certain other scientific developments, the new cosmology has fostered a major movement of reconciliation between the scientific and the religious communities. Book titles such as "God and the New Physics" (by physicist Paul Davis) or "God and the Astronomers" (by the astronomer Robert Jastrow) have come to abound, and it is hardly possible, these days, to keep up with the profusion of seminars and symposia on "science and religion" being held all over the world. And everywhere one encounters the same message of "peace and harmony" from the former contestants. There is however a price to be paid on the part of religion: wherever a conflict does arise—as between Genesis and the big bang—it is always Christianity which is obliged, by the presiding experts, to conform its teachings to the latest scientific theory. It appears that a certain fusion of science and religion is now in progress on a world-wide scale, which threatens to transform Christianity into some kind of "theistic evolutionism" more or less akin to the quasi-theology of Teilhard de Chardin.[10]

In a word, the new cosmology is not quite as innocuous as one might think. So far from being compatible with the truth of Christianity, it proves to be one of the most seductive and potentially lethal doctrines ever to threaten the integrity of the Christian faith: a dogma amply capable, it seems, of "deceiving even the elect." The devil, they say, gives us nine truths, only to catch us in the end with a lie; could big bang cosmology be a case in point? Could *this* be the underlying reason why an atheistic science has now promulgated—to everyone's amazement!—a doctrine which, on the face of it, glorifies God as the creator of the universe? It has at times been suggested that there is indeed a connection between the scientific enterprise and the demonic realm; this has been affirmed, for example, by the late Orthodox Hieromonk Seraphim Rose, and again by the Catholic historian Solange Hertz. It is not easy, of course, to document such a connection; but the surmise of demonic influence is neither irrational nor indeed improbable. When it comes to a major onslaught against the Catholic faith, it behooves us to recall the sobering admonition of St. Paul, which may well bear also upon the point at issue: "Put on the armour of God, that ye may be able to stand against the wiles of the devil. We wrestle not against flesh and blood, but against principalities, against powers, against the rulers of the darkness of the world, against spiritual wickedness in high places." (Eph. 6.11, 12) The demonic connection, then, of which we speak, may prove to be more than a pious fantasy. I find it remarkable that St. Padre Pio has referred to science as "the bible of Antichrist"; theologians can no doubt be fooled, but it appears that saints are not easily deceived.

Normal physics, one can say, has to do primarily with operational truth. Physics, after all, is indeed "the science of measurement," as I have pointed out before[11]; and so, in normal physics, ontological interpretation is optional: a private matter, relating to the philosophic orientation of the individual scientist. In the astrophysical domain, on the other hand, this no longer holds true, for the simple reason that astrophysics is virtually bereft of operational significance. The astrophysicist makes few if any predictions, and when he does predict, he typically misses the target by an order of magnitude or two. The object of astrophysics, it seems, is not primarily to predict phenomena, but somehow to construct a cosmology which is not too blatantly opposed to the observable facts. It is a new kind of physics, one which differs radically from the kind described in textbooks on scientific method. Now, a natural science which is not operational in its rationale can only be ontological in its claims, which is to say that in the case of astrophysics ontological interpretation is no longer optional but primary. In this regard physics "in the large" differs sharply from physics "in the small": from quantum theory, that is, where operational concerns are paramount. When Niels Bohr declared "there is no quantum world," this contention was neither inconsistent with the principles of quantum theory, nor was it abhorrent to the physics community at large; but imagine what would be the reaction if a scientist were to declare "there is no astrophysical universe"!

The ontology of astrophysics is of course physical, which is to say that one conceives of stellar objects as aggregates of fundamental particles. But why should this ontology be correct in the distant reaches of space-time when it fails (as I have argued repeatedly) in the terrestrial domain? If corporeal objects pertaining to the terrestrial mesocosm prove to be "more" than atomic aggregates, why should stars and galaxies be "nothing but" atomic? Moreover, if quantum particles "here below" do not have an independent existence—if they belong to what John Wheeler terms "the participatory universe"—why should it be otherwise in outer space? My initial contention is that a strictly physical ontology is as fallacious in the stellar domain as it is in the terrestrial sphere of perceptible objects.

I have elsewhere[12] characterized the knowledge of modern physics as "basic but inessential": basic, because it refers to the material side of cosmic reality, and inessential, because it is incapable of comprehending substantial form. Physics as such cannot know the quiddity or "whatness" of a thing; the very essence of things eludes its grasp.[13] Now, if to know a thing is to know its substantial form—to know, in other words, "what" the thing is—then it follows that the knowledge of physics is not a true knowledge. On the other hand, so long as physics remains closely tied to the observational domain—so long as its theories can indeed be tested by experiment or verified in their technological applications—it does evidently embody knowledge of a kind: a pragmatic or operational knowledge, as we have said. The problem, however, is this: Man was made to know, not pragmatically, but in truth. Hence the well nigh irresistible tendency to reify the intentional objects of physics by attributing to them a corporeal form. Basically, one treats the object in question as if it *could* be seen, could be touched. A fundamental particle, thus, becomes a tiny spherical ball, or perhaps some kind of wave which the mathematically trained can picture. I have pointed out that such processes of visualization play a necessary and indeed legitimate role in the comprehension of mathematical ideas even on the most abstract and sophisticated level: the human mind simply cannot do without some sensory support. It is in the domain of physics, however, as opposed to pure mathematics, that this art goes astray; whereas the mathematician understands full well the ontological difference between a function and its graphical representation, for example, the analogous distinction in physics is typically blurred. The reason for this confusion lies no doubt in the fact that the intentional objects of physics are admittedly more than an *ens rationis*, a mere "thing of the mind," even though they are *less* than a corporeal entity, less than a perceptible thing. And clearly, this circumstance imposes demands upon the ontological discernment of the physicist that are not easily met, to say the least. It thus comes about that the phantasmata of sensory representation are routinely projected upon the physical universe, which then becomes quite literally a fantasy world.

The real world can only be known by way of substantial form, by way of essence; but how can such knowledge be achieved? Strange as it may seem to the modern mind, we can and do know substantial form in the case of familiar corporeal entities: we know it through cognitive sense perception, to be exact. No use trying to explain such perception in terms of a natural process of whatever kind; as Whitehead says: "Knowledge is ultimate." Here below, "to know" and "to be" are mutually irreducible; and so we find that cognitive sense perception constitutes indeed a mysterious act, an act which transcends the confines of the natural world. As men of wisdom have pointed out: the eye by which we see the things of this world is not itself seen. What is seen by way of cognitive perception, let me recall, are corporeal entities; or more precisely: cognitive sense perception *gives access* to the corporeal domain.[14]

A major question presents itself: what about stellar objects? Can we also know *their* substantial forms, their very essence? Can we in fact perceive objects of that kind: do our powers of cognitive perception reach that far? When we see a dot of light in the night sky, are we actually perceiving a star or a galaxy? It does not seem that we are. What we perceive is a dot of light, which we may think of in generic terms as a "star." But a star in that sense is precisely something far away and high above, something which categorically exceeds our reach. Actual cognitive perception—the kind which takes place in the terrestrial realm and transcends bifurcation—does not presumably occur with reference to the stellar domain. Whatever may be the essence of a star or galaxy, that essence is not knowable by way of human perception; the stars, it turns out, are "above us" not only in a spatial, but also in an ontologic sense. I will point out in passing that this fact itself justifies and indeed entails the geocentrist claim.[15]

It appears that the aforesaid ontological recognition is indigenous to all the ancient cosmologies. Even St. Thomas Aquinas speaks of stellar substance as "incorruptible," and thus places stellar objects above the category of corporeal entities such as we find in the terrestrial domain. It is not mere poetry when St. Paul distinguishes between the two realms in 1 Corinthians 15: "There are also celestial bodies, and bodies terrestrial: but the glory of the celestial is one, and the glory of the terrestrial is another. There is one glory of the Sun, and another glory of the Moon, and another glory of the stars: for one star differeth from another in glory." Given that the intent of this discourse is to distinguish between the corruptible and the incorruptible bodies of man, it is clear that St. Paul is speaking in ontological terms. There is an implicit proportionality here: Celestial bodies are to the terrestrial as the resurrection body is to the natural. And to be sure, these two latter kinds of body are ontologically distinct; the second is "sown in corruption," the first is "raised in incorruption": there is literally a world of difference between the two. It would therefore be incongruous to suppose that there is no ontological hiatus between the stellar and the terrestrial realms to correspond to that difference; to think thus would be to reduce cosmic symbolism to a matter of appearance, a mere show. But such a trivialization is irreconcilably opposed to the Platonist ontology, and indeed to the *sophia perennis* in all its forms.[16]

Among the things in the natural world, the night sky, above all, speaks to us of high and sacred mysteries. Enclosing our earthly realm like an encompassing sphere, it awakens in us a sense of transcendence, an intimation of higher worlds. According to ancient belief, the starlight we see has its source in these higher worlds. The stars serve thus as a conduit, so to speak, an aperture in the vault of heaven through which the transcendent light breaks through to illumine the darkness of this nether realm. That celestial light, moreover, illumines not only the external world, but first of all the heart, the intellect of man. I should point out that there is a biblical basis for these ancient beliefs. To begin with, it is profoundly significant that Genesis refers to stars as "lights in the firmament of heaven," suggesting that the quiddity or essence of a star is indeed none other than "light." That stellar light, however, proves not to be primary, even in the order of creation, for we are told that the primary light was created on the first Day. Or better said: It is by virtue of that first-created light that the first Day itself is defined: one must remember that the first creative utterance of God was indeed the *Fiat lux*. The primary light, moreover, constitutes not only the first, but in fact the highest, the most godlike element in creation; it has been rightly called the most direct manifestation of God. And so it is also the highest symbol of God, a symbol hallowed by St. John when he declared that "God is light, and in Him is no darkness at all." (1 John 1.5) The first-created light, however, is not manifest in our world: as the source of all visibility, it exists unseen. Plato implicitly compares that unseen light to the light of the Sun when he refers to the latter as "the author not only of visibility in all visible things, but of generation and nourishment and growth." (*Republic*, Book VI) One is strongly reminded of Psalm 35: "For with thee is the fountain of life: in thy light shall we see light"! An entire metaphysics of light is concealed in either passage, a metaphysics which the Neo-Platonists were eager to unfold. And let us not forget that this doctrine, which is as biblical as it is Platonist, was incorporated into Christianity, notably through the teachings of St. Augustine and the Pseudo-Areopagite.

Getting back to the stellar realm, one sees that it constitutes a world of secondary yet supra-physical light, within which the primary light is mysteriously enshrined.[17] One might add that there is a profound connection between the stellar and the angelic realms, a matter which however would take us too far afield. Suffice it to say that a star is incomparably more than a mere aggregate of quantum particles, that it has both an essence and a function which vastly transcend the astrophysical domain. It is equally important, however, to recall that the stars were given to mankind "for signs." Admittedly this biblical affirmation may have an esoteric sense, by which I mean that there may have been a time when men were able to read "what is written in the stars." But it has also, most assuredly, a significance which applies to us all: for as I have noted before, the night sky awakens in us a sense of transcendence, a presentiment of celestial spheres. Even Immanual Kant, worlds removed, as he was, from the sapiential traditions, still sensed the grandeur of this cosmic icon; there are two things, he said, which fill the mind with wonder: "the star-spangled sky above me, and the moral sense within me." How strange that even this prosaic rationalist, whose philosophy is irreconcilably opposed to the *sophia perennis*, could still sense, however dimly, a connection between "the star-spangled sky" above and the "moral law" deep in the heart of man.

It is needless to say that this connection has disappeared—has been implicitly denied—in the astrophysical doctrine. The cosmic icon, set up by the Hand of God, has been broken; or better said, replaced by a man-made picture. Yes, the astrophysical panorama *is* a picture, a kind of image, and nothing more; whatever substance, whatever truth there be in contemporary astrophysics, must reside, as we have seen, in a mathematical structure and its operational interpretation, which is something else entirely. What astrophysics has to offer the general public is quite literally a fantasy world, a kind of science fiction cosmos; and it is this that has come in effect to replace the cosmic icon in educated minds.

I contend, moreover, that the astrophysical picture is seriously flawed, that it is in fact false. It is false, first of all, because it reifies the physical object. As I have explained at length in my monograph,[18] this is the root error of the contemporary *Weltanschauung*. It is the cause of "quantum paradox" in the microworld, and of a false reductionism in the corporeal domain: the hallowed belief that corporeal objects are "made of atoms." In the astrophysical realm, however, this generic error is compounded by the fact that stellar objects are not, strictly speaking, corporeal, that is to say, perceptible. Whereas in the mesocosmic or subcorporeal domain the reification of the physical entails a single error—a false reductionism, namely—in the astrophysical domain it entails *two*: for here one not only corporealizes the physical, but spuriously corporealizes the stellar substances as well. One thus destroys the dimension of transcendence, the verticality of "the above." The celestial is reduced to the terrestrial; the cosmos is homogenized—"democratized," one could almost say.

In thus distinguishing categorically between stellar and corporeal substances, I am adhering to the conception which associates corporeality with cognitive sense perception. It is on this basis, let us recall, that I differentiate between the physical and the corporeal domains. However, one can also speak of corporeality in a wider sense, corresponding to the Vedantic notion of gross (*sthûla*) manifestation. Corporeality in that sense is characterized by the conditions of space and time, and thus includes the stellar world in its spatio-temporal extension. The kind of corporeality to which cognitive sense perception gives access can now be qualified as terrestrial corporeality, the point being that there are modes of corporeality which differ from the terrestrial not just in terms of quantitative or measurable parameters, but in terms of essence.

The thesis of ontological heterogeneity might be contested on the grounds that the universe seems indeed to be very much of one piece: does one not detect the very same spectra of hydrogen, of helium, and of other elements, in the laboratory and in the light of stars and galaxies millions of light-years away? Yes, indeed one does; but we must realize that this kind of homogeneity pertains precisely to the physical realm, which stands below the level of being, of actual substance. What the physicist perceives, as it were, at the end of his analysis, are aggregates of quantum particles and nothing more; all ontological distinctions are thus obliterated. But what *are* these so-called particles to which everything has been reduced? As Heisenberg has put it, they are "a strange kind of physical entity just in the middle between possibility and reality." The problem with the homogeneous universe of the physicist is that it does not actually exist. At the risk of digression, I would point out that there is a lesson to be learned from this, one that applies even in the political and sociological domains: Obliterate ontological distinctions—obliterate *hierarchy*—and nothing at all remains; in a word, ontological homogeneity is tantamount to non-existence. But let us get back to the physical universe: At the end of the physicist's analysis, what remains is not one substance, but no substance at all. As Eddington has pointed out, the very idea of substance has no more place in physics—so long, of course, as that discipline is conceived rigorously, that is to say, in its mathematical structures and operational definitions. Thus, when a physicist fails to perceive a categorical difference between the substance of a star and the substance of a terrestrial entity, it is because, strictly speaking, he does not perceive any substance at all. Do not, therefore, ask an astrophysicist "What is a star?": in his capacity as astrophysicist, he has not the ghost of an idea.

One might add that what I have said with reference to stellar substances applies in principle to planetary bodies as well, beginning with the Moon. To be sure, men have walked on its surface and have brought rock samples back to Earth to be analysed; and yet I claim, in light of tradition, that lunar substance differs from terrestrial. Two issues are involved: first, the matter of physical inspection, and secondly, the new factor of close-up cognitive sense perception. As concerns the first, the preceding observations have made it plain that nothing new, nothing "non-terrestrial," can emerge from an inquiry of that nature: what we find, once again, are aggregates of quantum particles and nothing more. The fact that actual rock samples are now available for chemical analysis does not change the picture: the preceding considerations apply unaltered to this scenario as well. The matter of cognitive sense perception, on the other hand, is not quite so simple, and demands considerations of a very different kind. One needs to recall, in particular, what I have said in the Introduction to this book regarding experiential knowledge of cosmic realities: according to traditional doctrine, a stratum of cosmic reality can be "entered" only by actualizing the corresponding state in ourselves. So long, therefore, as we remain confined to the state corresponding to the terrestrial domain, terrestrial reality is all that we can perceive. To the extent that we are able to perceive lunar substances at all, we are bound therefore to perceive them as terrestrial, which is to say that in fact we do *not* perceive them. Take an animal—or for that matter, a man bereft of culture—into an art museum, and what do they see? What they see inside the museum is basically the same as what they see everywhere else: what is above that common level is not perceived. Such considerations, of course, do not prove the traditional claims regarding supra-terrestrial substances; they do suffice, however, to deflate the argument of those who maintain that these claims have now been disproved.

It is no doubt true that of all planetary bodies known to us, the Earth alone offers physical conditions capable of sustaining human life. To be sure, from a scientific point of view the physical environments associated with planetary bodies can presumably be explained in terms familiar to us all; and yet, from a traditional point of vantage, the matter is seen in an entirely different light. The prime determinant, now, becomes essence, the quiddity or inner content of these various bodies; the physical conditions, it turns out, are neither primary nor accidental, but are linked to essences. We understand this fact well enough when it comes to a living organism, the contours and physical characteristics of which are naturally expressive of its species; but even here we believe in the primacy of physical explanation, which is the reason why we are committed to an evolutionist biology. In a de-essentialized cosmos physical parameters are all that is left. When essences come into play, on the other hand, it becomes possible to understand the recognizable facts in an altogether different way, through what could be termed a "top-down" approach to cosmic reality. On this basis it becomes clear, in particular, that mankind finds itself on Earth, not on account of some physical contingency, but by virtue of a profound kinship. As I have noted before, it is on account of this inner kinship that we are able to "enter" the terrestrial stratum of cosmic reality by way of cognitive sense perception: the miracle of perception, I say, hinges upon a conformity of essence. As the matter stands, however, there is no such kinship in relation to the Moon or to Mars; and I would add that it is not an accident, therefore, that even the physical ambience of these planetary bodies proves hostile to man. Strictly speaking, contemporary cosmology is misnamed, because in truth it knows nothing of a cosmos, that is to say, of an *ordered* world.

Getting back to what might properly be termed global cosmography, I would note that the quantitative immensities of the stellar world, as documented by contemporary science, may conceivably be factual, hypothetical though they be. Let us suppose that they are: there remains the problem of their human assimilation. And this is ultimately what determines whether the immensities in question will prove enlightening, or whether in the end they will blind us and blight our humanity. I contend that only a true metaphysics—a metaphysics which is profoundly theological—can save us from the latter fate. "The heavens declare the glory of God, and the firmament sheweth his handiwork": only on that basis, I say, can we bear the immensities of the stellar world. It behooves us to realize that the heavens we perceive, be it directly or with the aid of telescopes, exemplify the heavens which we do *not* perceive, and that the quantitative vastness of the stellar universe mirrors the true immensity of the spiritual world. It is not a question here of symbolism in the anaemic sense of "the merely symbolic," but in the Platonist sense, rather, of actual "participation." The stellar universe, thus, "participates" in the metacosm, in the spiritual world, and it is this *ontological* fact that bestows a superior dignity upon the stars, a sacredness, one can almost say, which man is obligated to respect.

To the astrophysicist, on the other hand, a star is simply "a very hot gas," and nothing more. I have argued that such a reduction is epistemologically unfounded and metaphysically untenable; it remains to comment on its effect upon our humanity. From a traditional point of vantage, to be sure, the astrophysical reduction is a profanation, a kind of sacrilege; but what impact does it have upon an already profane civilization? Does a cosmic symbolism retain any kind of efficacy when it is no longer recognized, no longer understood? I surmise that the efficacy of an authentic symbol survives its comprehension: symbols do not die. The stellar universe, I maintain, retains a paramount iconic significance even in the present iconoclastic age: it is only that its significance has become inverted (here it is again: the diabolic connection!). What we think about the stars, how we picture the stellar world, does still have its effect on us; whether we realize it or not, it does influence and profoundly affect our views regarding God, man, and human destiny. The heavens, I contend, will declare, either "the glory of God," or the supreme futility of existence: here there can be no middle ground, precisely because the stellar world, in its iconic function, signifies the highest cosmic sphere. If that consists simply of particles engaged in meaningless motion, then all human aspirations must in the end prove vain. If the stellar light, which the ancients thought to be of celestial origin and which Plato viewed as the carrier of intelligible essences—if that light fails, the cosmos and all that it contains is reduced in the final count to nothingness. It is surely no accident that the rise of astrophysics has been accompanied by the advent of postmodernist nihilism, in its philosophic as well as its cultural manifestations. The drift into nihilism corresponds precisely to the loss of substance implicit in the physicist's worldview; culture and cosmology, it turns out, are intimately connected, and it appears that when the prevailing cosmology flattens, so does the culture.

To conclude: The hierarchic distinction between stellar and terrestrial substances is vital to a sound cosmology. What I have previously termed the rediscovery of the corporeal needs therefore to be followed by another basic recognition: the rediscovery of the stellar world.

Notes

1. Georges Lemaître, a student of Arthur Eddington, was a Belgian Jesuit and physicist. His speculations concerning "the primeval atom"—a curious blend of physics and philosophy—seem to have been well received in ecclesiastic circles, judging by the fact that he was soon thereafter appointed director of the Pontifical Academy of Science. The idea of the big bang, oddly enough, goes back to the poet Edgar Allen Poe, who was also an avid science amateur. To counter the problem of gravitational collapse, he proposed, in 1849, that the universe came to birth in an explosion.
2. Like Stephen Hawking in *A Brief History of Time*, Gamov commended his vision of the universe to an entire generation in a book entitled *One, Two, Three, Infinity*, which became a science best-seller.
3. For instance, when astronomers discovered discrepancies between observed planetary orbits and the trajectories predicted on the basis of Newtonian physics, they conjectured that these deviations may be due to an as yet unidentified object. This hypothesis was verified in 1930 with the discovery of the planet Pluto.
4. It can be explained, for example, by the quasi-steady-state theory of Burbidge, Hoyle and Narlikar (see *Physics Today*, Vol. 52, No. 4, April 1999, 38-44), or by the plasma-physics approach of Hannes Alfvén (*Cosmic Plasma*, Holland: D. Reidel, 1981), both of which appear to be viable alternatives to big bang cosmology.
5. Halton Arp, *Seeing Red: Redshifts, Cosmology and Academic Science* (Montreal: Apeiron, 1998), 95. Halton Arp is an American astrophysicist, one of the foremost authorities on quasars. However, when he began openly to question the Doppler interpretation of stellar redshifts, he quickly became *persona non grata*, at least in the United States. He is now at the Max Plank Institut für Astrophysik at Munich. His book constitutes a valuable resource in a field in which it is becoming increasingly difficult to separate fact from fiction.
6. Ibid., 225-233.
7. Foremost among this handful of dissenting scientists is the late Fred Hoyle, one of the original pioneers of astrophysics.

8. As we shall presently see, however, it is equally significant that Pope John Paul II, in another address to the Pontifical Academy, delivered in 1988, warned against "making uncritical and overhasty use for apologetic purposes of such recent theories as that of the big bang."

9. Henry Denzinger, *The Sources of Catholic Dogma* (London: Herder, 1957), 2121-2128. It is to be noted that Pope St. Pius X, in his *Motu proprio* of 1907, "Prestantia Scripturae," has declared the decision of the Biblical Commission to be binding. See Denzinger, 2113.

10. On the question of Teilhardism, I refer to my monograph *Teilhardism and the New Religion* (Rockford, IL: TAN Books, 1988).

11. "Eddington and the Primacy of the Corporeal," *Sophia*, Vol. 6, No. 2, 2000, 5-38, reprinted here as Chapter 3.

12. "*Sophia Perennis* and Modern Science," in *The Philosophy of Seyyed Hossein Nasr*, The Library of Living Philosophers, Vol. XXVIII (2001), 469-485, reprinted here as Chapter 1.

13. I might note that Jacques Maritain has said much the same when he characterized modern physics as "perinoetic."

14. I have dealt with this question in "Eddington and the Primacy of the Corporeal," op. cit., 24-27. See Chapter 3, 61-63.

15. On the question of geocentrism, I refer to Chapters 8 & 9.

16. A lion, for example, according to ancient belief, not only symbolizes the Sun by virtue of its mane, but is in fact a "solar" beast. But of course this is something modern man has ceased to understand: his diminished ontology admits no such connections. For that one needs an ontology which is open to what may properly be termed "the mystery of essence."

17. As Arthur Zajonc, an authority on quantum physics, has beautifully said of light as such: "I cannot describe it, my imagination can only just touch its hem, but I know that at its core there seems to live an original 'first light' within which wisdom dwells, a wisdom warmed by love and activated by life" (*Catching the Light*, Oxford University Press, 1995, 325). I find it truly remarkable that a contemporary physicist should bear witness to the perennial metaphysics of light, and in such eloquent terms, no less. Like the iconic bright spot within the dark field, Zajonc compensates, as it were, for the almost universal nescience of his scientific peers.

18. *The Quantum Enigma*, (Peru, IL: Sherwood Sugden, 1995).

Chapter VIII

The Status of Geocentrism

If there has been little debate in recent times on the subject of geocentrism, the reason is clear: almost everyone takes it for granted that the geocentrist claim is a dead issue, on a par, let us say, with the flat-Earth hypothesis. To be sure, the ancient doctrine has yet a few devoted advocates in Europe and America, whose arguments are neither trivial nor uninformed; the problem is that hardly anyone else seems to care, hardly anyone is listening. Even the biblically oriented creation-science movement, which of late has gained a certain prestige and influence, has for the most part disavowed geocentrism. The fact remains, however, that geocentrist cosmology constitutes not only an ancient, but indeed a traditional doctrine; should we not presume that as such it enshrines a perennial truth? To maintain, moreover, that this truth has nothing to say on a cosmographic plane—that the doctrine, in other words, is "merely symbolic or allegorical"—to think thus is to join the tribe of theologians who are ever willing to "demythologize" at the latest behest of the scientific establishment. It will not be without interest, therefore, to investigate whether the geocentrist claim—yes, understood cosmographically!—has indeed been ruled out of court. I shall argue that it has not. As regards the Galileo controversy, I propose to show that Galilean heliocentrism has proved to be scientifically untenable, and that in fact the palm of victory belongs to the wise and saintly Cardinal Bellarmine. I should add that the problematic of this article will lead us, in the final section, to elicit an interpretation of relativistic physics that accords with traditional doctrine.

Weighing the different systems. Giovanni Battista Riccioli's (an adaptation of Tycho Brahe's model) versus that of Copernicus. From Riccioli's Almagestum Novum.

Nothing perhaps is more impenetrable to the modern mind than the ancient cosmologies. Not only are these cosmologies inherently metaphysical, but what perhaps we find more puzzling still is the fact that they are also at the same time descriptive of the perceptible world, at least in a qualitative sense. When such cosmologies speak of the Sun, Moon, and stars, the reference is no doubt symbolic, but not "merely symbolic," which is to say that the cosmologies in question are not bereft of scientific content. It is this fusion of a "metaphysics" with a "physics," a natural science, that seems to baffle us the most. It needs of course to be understood that the kind of science to which I allude is worlds removed from the Baconian enterprise, both in point of method as well as in the end to be achieved. In particular, it appears that ancient cosmologists did not feel obliged to account for such things as retrograde planetary motions, or the precession of equinoxes; according to Thomas Kuhn, "Only in our own Western civilization has the explanation of such details been considered a function of cosmology. No other civilization, ancient or modern, has made a similar demand."[1] Be that as it may, a shift from cosmology of the ancient kind to astronomy in the modern sense can be clearly discerned in Greece from the time of Pythagoras to Ptolemy. One has the impression that mythical and inherently metaphysical notions, such as that of the heavenly "spheres," were being gradually transformed into physical conceptions, destined to be scorned and discarded with the advent of modern times. Meanwhile mathematical techniques of increasing complexity and sophistication were devised to attain ever greater accuracy in the description of observable phenomena.

The simplest part of Greek astronomy pertains to the stellar sphere. It turns out that stellar orbits could be described, with what seemed at the time to be perfect accuracy, on the assumption that the stars are fixed on the surface of an immense sphere concentric with the Earth, which rotates diurnally around an axis that could be identified within one degree by the position of the North Star, while the Earth remains stationary at the center of the universe. If it were not for the Sun and other "wanderers," the simple "two-sphere" model of ancient astronomy would have provided a seemingly perfect description of the relevant phenomena. But the orbits of the so-called planets proved to be complex and challenging in the extreme. Eventually Greek astronomers concluded that circular orbits concentric with the terrestrial sphere will not suffice for their description, and by the time of Hipparchus, whose active life can be dated between 160 and 127 B.C., the method of epicycles and deferents had come into use. The motion of a planet was now conceived as the sum of two circular motions: a small rotation, namely, around a center known as the deferent, which itself sweeps out a much larger circle centered upon the Earth. The epicycles were thus conceived as a small correction to the circular orbits of the older "spherical" astronomy. I would add that in addition to his work with epicycles, Hipparchus is said to have discovered the precession of the equinoxes, which he estimated at 36 seconds of arc per year (as compared to its actual value of some 50 seconds). He also estimated the distance to the Moon to be 33 times the diameter of the Earth, as compared to an actual value near 30.2. Following upon these discoveries, progress in planetary astronomy continued apace. It was not long before astronomers realized that orbits could be progressively corrected by adding epicycles to epicycles in an indefinite series. Moreover, they discovered that additional corrections could be introduced by displacing the center of a deferent. The resultant circles, known as eccentrics, could be adjusted to further improve the result. In addition, it was found that a higher degree of accuracy could also be achieved by taking the rate of rotation of a deferent or some other point in the geometric scheme to be uniform, not with respect to its actual center, but with respect to a displaced center known as the equant. The method of epicycles came thus to be supplemented by the use of eccentrics and equants. One can say in retrospect that Greek astronomy furnishes the first example of mathematical modeling on a serious scale. The development culminated in the work of Claudius Ptolemy, whose treatise known as the *Almagest* (circa 150 A.D.) dominated Western astronomy till at least 1543, when it began to be displaced by the Copernican theory.

What motivated Copernicus to reject the Ptolemaic theory in favor of a heliocentric astronomy? In his preface to the *De Revolutionibus*, Copernicus cites persistent inaccuracy and lack of coherence as his principal criticism of the prevailing astronomy. Some fourteen hundred years after the publication of the *Almagest*, the computational problems of planetary astronomy had not yet been solved with satisfactory precision. Worse still, there seemed to be no principle, no rhyme or reason, governing the proliferation of epicycles, eccentrics and equants. All these mathematical parts and pieces did not seem to fit together into a coherent whole. "It is as though an artist were to gather the hands, feet, head and other members for his images from diverse models," writes the Polish astronomer, "each part excellently drawn, but not related to a single body, and since they in no way match each other, the result would be monster rather than man." It has been pointed out that this perception reflects the Neoplatonist influences to which Copernicus was demonstrably exposed. In any case, what renders the expanded Ptolemaic model monstrous in the eyes of Copernicus is the *ad hoc* character of its multiple constructions, which is to say, its lack of mathematical intelligibility as a whole. By way of contrast, he maintains that the new heliocentric astronomy exhibits "an admirable symmetry" and "a clear bond of harmony in the motion and magnitude of the spheres." Consider, for example, the retrograde motion of planets: why should a planet reverse its normal eastward motion and retrogress for a time, till it resumes its eastward course? From the standpoint of Ptolemaic astronomy, this phenomenon presents itself as an inexplicable irregularity, which can indeed by accounted for through the introduction of appropriate epicycles, but can hardly be understood. From a heliocentric point of view, on the other hand, retrograde motion ceases to be an irregularity, and becomes simply a mathematical consequence of the fact that the Earth itself is in motion around the Sun. It is easy to see that this is the case. Simply draw three concentric circles representing the sphere of the stars, the orbit of the Earth, and the orbit of the planet. In the case of an inferior planet (Mercury or Venus), the latter circle will be the innermost, whereas, for a superior planet, the orbit of the Earth will be contained within the other two. If now we mark successive positions of Earth and planet on their respective circles[2] and connect corresponding points by a line to obtain "observed" planetary positions on the stellar sphere, we can readily see how the motion of the Earth gives rise to the phenomenon of retrogression. What is more, we discover that retrogression occurs when the planet is nearest to the Earth, which explains why retrogressing planets are observed to shine more brightly. Copernican astronomy, it appears, has provided the first scientific explanation of the fact that planets retrogress.

A second irregularity which the new theory explains beautifully is the observed variation in the periods of orbital planetary motion. Given that the Earth itself orbits around the Sun, the time it takes for a planet to return to its starting position, as observed from the Earth, is clearly not the same as the time it takes to complete an orbit. The observed variations in planetary orbital periods can therefore be explained, once again, by the postulated motion of the Earth. Copernicus cites many examples of this kind to document the explanatory power of the heliocentric hypothesis. I will mention one more: It is known that the planets Mercury and Venus can only be observed in the vicinity of the Sun. To account for this fact in Ptolemaic terms, it was necessary to introduce deferents and epicycles binding these planets to the Sun. Heliocentric astronomy, on the other hand, requires no such *ad hoc* constructs: the given phenomenon is an immediate consequence of the fact that the orbits of Mercury and Venus are contained *within* the orbit of the Earth.

This brings us to a major point of difference between Ptolemaic and Copernican astronomy: In the heliocentric scheme, the order of the planetary orbits can be determined from observational data. If the planets (including the Earth) revolve in circular orbits around the Sun, it is possible in fact to calculate the ratios of the planetary radii in terms of the angular distances from the Sun to the planets as measured from the Earth. Such is not the case in a geocentric system, where not even the order of the planets is determined by the appearances. In a word, the new astronomy is far more coherent than the old. It is this newly-discovered coherence that Copernicus is alluding to when he speaks of "a clear bond of harmony in the motion and magnitude of the spheres," and it is evident that he views this new "bond" as a powerful argument for the truth of his theory.

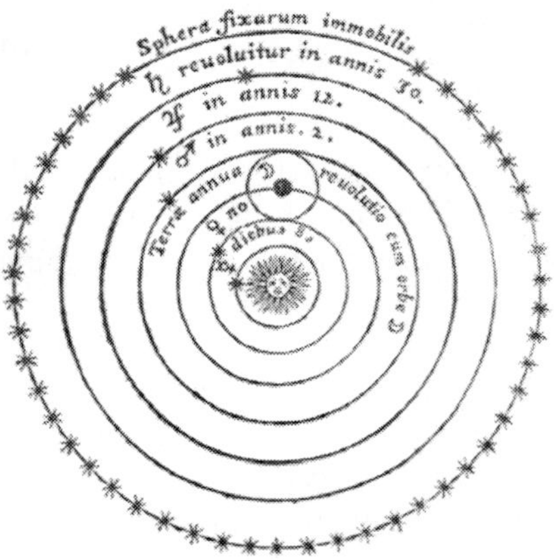

The heliocentric system of Copernicus.

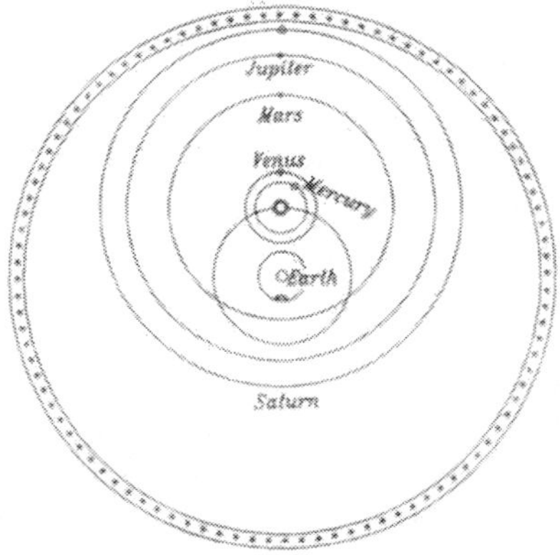

The geo-heliocentric system of Tycho Brahe.

In actuality, however, the charge of incoherence applies to the new astronomy as well. In practice Copernicus was forced to introduce epicycles and eccentrics of his own, and like Ptolemy himself, ended up with over thirty circles, without any appreciable gain in the degree of accuracy. Clearly, the problem of planetary astronomy had not yet found its solution.

It appears that the decades following publication of the *De Revolutionibus* witnessed few converts to the Copernican cause. To the general public the notion of an orbiting Earth seemed both absurd and impious, and even astronomers seem for the most part to have been wary of that hypothesis. The second half of the century, moreover, was dominated by the imposing figure of Tycho Brahe, a powerful opponent of heliocentrism. Brahe is known, first of all, for the uncanny accuracy of his astronomical measurements. His results are frequently precise to one minute of arc, an unrivaled achievement for naked-eye observation. What especially concerns us, however, is the fact that Brahe proposed a remarkable planetary theory of his own, which to this day finds partisans in Europe and America.[3] Accepting the traditional notion of an immobile Earth and a stellar sphere engaged in diurnal geocentric rotation, he proposed that Mercury, Venus, Mars, Jupiter, and Saturn circle the Sun, while the Sun and Moon circle the Earth. It happens that this geocentric theory embodies all the advantages previously cited by Copernicus in behalf of his heliocentric model. It too explains such things as retrograde motion, the variation of planetary periods, and the binding of inferior planets to the Sun, without recourse to epicycles or other *ad hoc* constructions. In point of fact, it can be shown that the two planetary theories—the Tychonian and the Copernican—are mathematically equivalent, which is to say that they predict exactly the same apparent planetary trajectories.[4]

It should however be noted that the two theories are not equivalent in regard to stellar astronomy, for it is evident that a displacement of the Earth would entail a corresponding parallactic shift in the apparent position of a star. Tycho Brahe himself had searched for such a shift, but found none. This means that stellar parallax, if it exists, must be of an order of magnitude less than a minute of arc, which would necessitate stellar distances far greater than astronomers were wont to assume. Copernicus himself had recognized that his hypothesis demands an enormous enlargement of the stellar sphere, and it may be worth noting that Tycho Brahe considered this fantastic multiplication of apparently empty space to be one of the most cogent reasons for rejecting the Copernican hypothesis. Yet, from a strictly scientific point of view, a decision between the two theories could not be made at the time.

Although the problem of planetary astronomy, as I have said, had not yet been solved, it turns out that both Copernicus and Tycho Brahe, each in his own way, had made decisive contributions which were soon to lead to a definitive solution. The breakthrough came in the first decade of the seventeenth century at the hands of Johannes Kepler. Availing himself of superior data supplied by Tycho Brahe, he proposed a new heliocentric theory which was destined to carry the field. After years of futile endeavor, Kepler abandoned the time-honored method of epicycles in favor of a radically new idea: he proposed that the planets revolve around the Sun in elliptical orbits with variable speed. His so-called First Law stipulates that the Sun is situated at one of the two foci of the planetary ellipse, while his Second Law states that the line segment from the Sun to the planet sweeps out equal areas within the ellipse in equal times. Qualitatively, this simply affirms that the planet moves faster the nearer it is to the Sun; in point of fact, however, Kepler's "law of equal areas" enables one to calculate the velocity of the planet at every position of its trajectory. In conjunction, the two laws lend themselves to a complete description of the planetary system. No need any longer for epicycles, eccentrics, equants, or other devices of the kind: it turns out that two simple and mathematically elegant laws suffice to solve the age-old problem. As Thomas Kuhn points out: "For the first time a single uncompounded geometric curve and a single speed law are sufficient for predictions of planetary position, and for the first time the predictions are as accurate as the observations."[5] It would take us too far afield to comment on the Neoplatonist basis of Kepler's conceptions; suffice it to say that he first recorded his new ideas in a treatise on the motion of Mars, the most challenging of the planets. It could well be said that the era of modern astronomy commences with the publication of this work, in the year 1609.

It was in the same year, 1609, that Galileo Galilei first turned his telescope to the sky, with startling results. In quick succession he discovered the Milky Way to be a sea of stars, detected mountains and craters on the Moon, the height and depth of which he could estimate from shadows, found dark spots on the Sun, showing that the Sun itself rotates around its own axis, and discovered that Jupiter has four moons. While none of these findings have a direct bearing on the Copernican question, they have had a decisive impact upon the European mentality in that they appeared to discredit the perennial distinction between the celestial spheres, which mankind had taken to be perfect and immutable, and the "sublunary" world: this imperfect and ever-changing domain which constitutes our habitat. Peering through his telescope, Galileo seemed to behold one and the same kind of world wherever he looked: from the rugged landscape of the Moon to the moving spots on the Sun itself. It is hard for us to imagine the sensation in European circles stirred by reports of these new vistas. A widespread fascination with astronomical discoveries seems to have gripped European society, eliciting varied reactions. John Donne—to cite perhaps the most striking example—appears to have sensed the deeper significance of the Galilean "movement" almost immediately: "And new philosophy calls all in doubt," he penned back in 1611; "'Tis all in pieces, all coherence gone." For the most part, to be sure, the response was less perceptive; as Kuhn points out on the lighter side, "The telescope became a popular toy." Yet far more than a toy! There can be no doubt that the new images gleaned through the telescope have contributed significantly to the demise of the ancient *Weltanschauung*.

One more Galilean discovery needs to be mentioned: the phases of Venus, namely, which seemed indeed to imply that Venus orbits around the Sun. But whereas Galileo exhibited this discovery as proof of the Copernican hypothesis, the fact remains that the phases of Venus are accounted for equally well on the basis of Tychonian astronomy. "It was not proof," writes Kuhn, "but propaganda."

Where, then, did the matter stand at the time of the Galileo controversy? One sees in retrospect that it stood very much as Cardinal Bellarmine had stated the case in his letter to Foscarini, in 1615: "To demonstrate that the appearances are saved by assuming the Sun at the center and the Earth in the heavens is not the same thing as to demonstrate that in fact the Sun is in the center and the Earth in the heavens," writes the Cardinal. "I believe the first demonstration may exist," he goes on to say, "but I have very grave doubts about the second . . ." Yes, the work of Johannes Kepler does indeed clinch the first demonstration alluded to; as regards the second, the subsequent history of science has fully justified St. Bellarmine's "grave doubts." As I propose to show presently, the science of our day has in fact rendered the second demonstration unthinkable.

With the publication of Newton's *Principia* in the year 1687 the scientific triumph of Keplerian astronomy was complete. Kepler's First and Second Laws could now be derived theoretically on the basis of a brilliant new physics, a physics that could be tested and verified in a thousand ways. So far as the scientific community was concerned, geocentrism was now a dead issue. No one doubted any longer that the Earth does move; it only remained to design experiments that could detect and measure that motion. What were these experiments, and what did they prove?

One avenue of approach relates to the phenomenon of aberration. In 1676, a Danish astronomer named Olaus Roemer noted that the period between observed eclipses of one of Jupiter's moons varies by several minutes, depending upon the relative position of the Earth. He concluded that light propagates at a finite velocity, which he estimated to be 309,000 kilometers per second (a result, one might add, which is accurate to within 3%). Now, if the Earth itself moves, that additional velocity will cause a shift in the apparent position of a celestial object. Think of a car driving through rain on a windless day. Relative to the car, the rain falls, not vertically, but at an angle, which depends upon the ratio of two velocities: the horizontal velocity of the car, namely, divided by the vertical velocity of the rain. From a measurement, therefore, of the so-called angle of aberration, one can determine the ratio of the velocities in question. This is the idea behind what appears to be the first experiment designed to demonstrate and measure the orbital velocity of the Earth. In 1724, James Bradley, the English Astronomer Royal, attached a telescope to the top of a chimney and began to observe the star Gamma Draconis, situated almost ninety degrees above the horizon. As expected, he found that in the course of a year the apparent position of the star described a small circle, corresponding to an angle of aberration close to 20 seconds of arc. By simple trigonometry, that angle equals the arctangent of v/c, where v denotes the orbital velocity of the Earth and c the speed of light. It follows that an aberration of 20 seconds corresponds to an orbital velocity close to 30 kilometers per second, in good agreement with astronomical theory *à la* Kepler and Newton. By observing and measuring the aberration caused by the orbital velocity v, Bradley had apparently observed and measured the stipulated motion of the Earth. Galileo's celebrated *Eppur Si Muove*, so it seemed, had at last been confirmed, vindicated before the world.

The story, however, does not end at that point. In 1871, another British astronomer, named George Biddel Airy, conducted an experiment based upon an idea which had been proposed more than a century earlier by a Jesuit named Boscovich. The latter had pointed out that if the telescope in Bradley's experiment had been filled with water in place of air, the resultant angle of aberration would have been increased, due to the fact that the velocity of light in water is less than in air. However, when the experiment was finally carried out, it was found that the angle in question had not changed at all! On the face of it, this result disproves the claim that Bradley's shift of 20 seconds is caused by aberration. To everyone's amazement, the argument *against* the motion of the Earth seemed now to be logically compelling: Given that orbital motion implies aberration, it follows indeed that the absence of aberration implies the absence of orbital motion. Understandably, the failure of Airy's experiment sent shock waves through the scientific community. However, worse was yet to come. In 1887, Michelson and Morley conducted their now famous experiment, designed to detect and measure the orbital velocity of the Earth, not by way of aberration, but by comparing the observed velocity of light in the direction of that orbital motion with its velocity in the opposite direction. According to elementary considerations, the two velocities should differ by exactly 2v, where v again denotes the orbital velocity. But again, to everyone's consternation, it turned out that the two light velocities are exactly the same, which is to say that the experiment yielded a measured orbital velocity equal to zero. Two crucial experiments, based upon different physical principles, had now reached the same conclusion: the Earth does not in fact move.

At this juncture one has only two options: one can accept the verdict that the Earth does not move and opt for a duly refined Tychonian astronomy, or else one can search for a way of adjusting the laws of physics so as to render the Earth's orbital velocity to be in fact undetectable. History records that Albert Einstein opted for the second alternative, and in so doing, astonished the world with his theory of relativity. The special theory ingeniously "explains" the negative results of both the Airy and the Michelson-Morley experiments, while, at the same time, it leads to a host of other testable predictions which have since been verified. Without question, Einsteinian physics constitutes one of the most brilliant and successful ventures in the history of modern science. One must not however forget that it operates by the logic of "saving appearances," which is not the same thing, to paraphrase Cardinal Bellarmine, as demonstrating that what is claimed in theory is in fact the case. As Walter van der Kamp—that indefatigable champion of Tychonian astronomy—was fond of pointing out, the logic of relativity constitutes a *ponendo ponens* argument: from the premise "P implies Q," one falsely concludes "If Q, then P."

Getting back to Bradley's experiment, the question remains: If indeed the Earth does *not* move, and there is consequently no stellar aberration, how is the small circle described by Gamma Draconis to be explained? On a Ptolemaic or Tychonian basis, the answer is clear: the observed phenomenon is caused by an actual circular motion of Gamma Draconis relative to the Stellatum, the revolving sphere of the stars. A similar remark applies to the phenomenon of stellar parallax, which was finally detected by Henderson in 1832, and measured with greater accuracy in 1838 by Bessel and Struve. From a geocentric point of view there is obviously no such thing as stellar parallax, which is to say that the phenomenon in question must be caused, once again, by corresponding stellar motions. These motions, to be sure, are too small to be observable to the naked eye, and for this reason were not known in ancient times. My point is that Henderson's discovery, so far from proving stellar parallax, can be accounted for on a geocentric basis as well.

This does not however imply that it is a matter of indifference to stellar astronomy whether or not we adopt a geocentrist point of view: nothing could be further from the truth! The fact is that in contemporary astronomy the hypothesis of parallax plays a crucial role in the determination of stellar distance. This can already be seen from the circumstance that the standard unit of astronomic distance is the parsec, which is defined to be the distance at which a stellar object subtends an angle of 1 second relative to a baseline whose length equals the mean distance from the center of the Earth to the center of the Sun. Clearly, the parsec is a parallactic unit of length. It means, in effect, that a star situated 1 parsec from the Earth will have parallax on the order of 1 second. It is to be noted that the hypothesis of parallax imposes enormous dimensions upon the stellar universe: a single parsec turns out to be about 31 million million kilometers. Moreover, since most stars—all but a relative few—have no measurable parallax, their distance from the Earth must be at least 25 to 50 parsecs; and in point of fact, stellar astronomers are wont to employ the megaparsec as their preferred unit of distance. A Tychonian universe, by comparison, would still need considerable size: but nothing like the billions of light years to which contemporary astronomy lays claim. One sees that a geocentric interpretation of astronomical phenomena, by eliminating stellar parallax, would undercut our current picture of the stellar world. A Tychonian universe would differ radically, both in size and in architecture.

But whereas contemporary astronomy is thus implacably opposed to the geocentrist hypothesis, it happens that pure physics is not. According to general relativity, it is in fact permissible to regard the Earth as a body at rest: as Fred Hoyle has put it, the resultant theory "is as good as any other, but not better." Relativity implies that the hypothesis of a static Earth is not incompatible with the laws of physics and cannot be experimentally disproved. To be sure, physics as such cannot affirm that hypothesis; but neither can it deny its validity. Already in 1904, Henri Poincaré had understood that "the laws of physical phenomena are such that we do not have and cannot have any means of discovering whether or not we are carried along in a uniform motion of translation"[6]; and by 1915, Einstein had concluded that the same applies to arbitrary motion. It appears that so far as physics is concerned, the geocentrist claim remains viable.

Given that the scientific challenge to geocentrism derives, not from physics, but from astronomy, we need to ask ourselves whether the latter science is in a position to prove its case. Our scientific knowledge concerning the stellar universe is of course based upon observations carried out either in terrestrial observatories or by means of instruments transported into outer space via satellites. It is crucial to note that the transition from observational data to claims concerning the stellar realms cannot be accomplished on the ground of physics alone, but requires additional hypotheses of an untestable kind. We have already encountered one such hypothesis in the assumption of stellar parallax, and the informed reader will likely recall the Doppler interpretation of stellar redshifts, as also the so-called cosmological principle, as further untestable assumptions basic to contemporary astronomy.[7] The logic, here, is once again of the *ponendo ponens* variety, which is to say that the hypotheses in question are judged or validated by their success in "explaining" observable phenomena. But apart from the circumstance that such an argument is never compelling—as Cardinal Bellarmine pointed out long ago!—it happens that the prevailing astrophysical cosmology has so far flunked the empirical test: despite inflated reports, and decades of concerted effort, one finds that big bang theory does not square with the empirical data.[8] Here too one has grounds to surmise, when the chips are down, that we are dealing, not indeed with "proof," but once again with "propaganda" of sorts. If the Airy or Michelson-Morley experiments had yielded their intended result, the scientific case against geocentrism, though still not compelling, would have been at least impressive; however, as the matter stands, the ancient doctrine has not even been rendered improbable, let alone has it been disqualified.

The late Walter van der Kamp has made an interesting point: "In truth," he said, "the choice is between Tycho Brahe and Einstein—Galileo *et al* are played out." Now, I certainly agree with the part about Galileo. To be precise, he is "played out," not because he affirmed the motion of the Earth, but because he staked the claim on ostensibly scientific grounds. One knows today on the basis of physics that Galileo's arguments are inconclusive, and that in fact the stipulated motion cannot be proved at all. What I find misleading, on the other hand, is the notion of "choosing" between Tycho Brahe and Einstein. I reject the implication that Tychonian astronomy and Einsteinian physics are mutually exclusive alternatives. I shall argue not only that the two positions are compatible, but that each has its own validity. The key to the problem lies once again in the discernment of what may be termed levels of cosmic reality: though interrelated, these levels remain distinct and need to be distinguished.

To begin with the physical, let me recall that I use this term in a technical sense: it refers to the level or aspect of cosmic reality which answers to the modus operandi of physical science. The physical is thus defined and known through acts of measurement, which entails that it owns neither essence nor substance in the traditional ontologic sense. As Heisenberg observed with reference to quantum particles, physical objects constitute a strange new kind of entity "just in the middle between possibility and reality."[9] Now, I will argue that on the level of these strange new entities, Einsteinian relativity has its place. It is by no means surprising, I say, that entities defined through acts of measurement should conform to principles of relativity: in the final analysis, are they not indeed relational by definition? Consider the case of velocity, the magnitude of motion: what else can it be, physically speaking, than a speed relative to this or that coordinate system, this or that frame of reference? On the physical plane there are indeed invariants—quantities, namely, which turn out to be the same relative to every coordinate system in some class—but no absolutes: nothing which is not inherently relational. For this one needs *essence*, something that answers to the question "What?". A stone, thus, or a cat is not relative, not simply relational, by virtue of the fact that it is something: a thing that is *more* than "midway between possibility and reality." A stone or a cat, therefore, is something which cannot be quantified, cannot be elicited by acts of measurement, and consequently does not pertain to the physical domain. As Eddington observed, "The concept of substance has disappeared from fundamental physics."[10] Strictly speaking, physics does not deal with *things*. It is true that the notion of substance was retained in what is now called classical physics; that retention, however, was spurious, that is to say, inconsistent with the operational principles of physical science. As Eddington points out: "Relativity theory made the first serious attempt to insist on dealing with the facts themselves. Previously scientists professed profound respect for the 'hard facts of observation'; but it had not occurred to them to ascertain what they were."[11] Albert Einstein was presumably the first physicist to arrive at a consistent theory based upon operational principles. It turns out that the tenets of relativity are not in fact *ad hoc* stipulations introduced to safeguard the Copernican hypothesis, as geocentrists are wont to claim, but respond to the very principles upon which physical science is based. It appears that in a world defined operationally—in what John Wheeler terms the participatory universe—Einsteinian relativity reigns supreme.

But how does the matter stand on the level of the corporeal world? Do the principles of relativity apply to a world that is to be known, not through acts of measurement, but through acts of cognitive sense perception? Do Einsteinian principles still apply in a world where essences manifest in the form of sensible qualities, a world in which not only relations but actual substances are to be found? There is no reason to believe that this is the case. Nothing obliges us to suppose that in this corporeal domain, which is distinctly "more" than the physical, there can be no absolute rest and absolute motion, nor absolute simultaneity of events.[12] If, on this higher level, stones and cats are real, why not also rest and motion, why not simultaneity of events as well? I have argued in *The Quantum Enigma* that it is the loss of essence, of substantial being, as one descends to the physical plane, that brings into play the quantum-mechanical superposition principle; I would like now to propose that the same loss, the same reduction, leads macroscopically to Einsteinian relativity. Where there is no substance to distinguish one reference frame from another, it is hardly surprising that the two are of equal weight. It is, as always, the loss of substance, of hierarchy in fact, that leads to the "democratization" of what remains.

One knows that the superposition principle is abrogated on the corporeal plane: one knows, for example, that a cat cannot be both dead and alive. It is this simple fact, as it turns out, that accounts for the so-called collapse of the state vector, which has mystified physicists since the advent of quantum theory in 1926.[13] Is it not reasonable, then, to suppose that the principles of relativity are likewise abrogated on the corporeal plane? Now, this abrogation, which I take to be factual, has obviously immense implications. It suggests, for example, that Aristotelian physics may not, after all, be quite as chimerical as we generally assume. Furthermore, the aforesaid recognition evidently casts a new light on the geocentrist controversy. One sees, in particular, that Tychonian astronomy may be more than a merely "admissible" theory, as general relativity declares: more indeed than simply "as good as any other, but not better." The point is that the worth of geocentric astronomy can no longer be ascertained exclusively through acts of measurement. The question, one finds, cannot be resolved on purely physical grounds, but calls for considerations of a different order. The most that physics as such can say is that geocentric astronomy cannot be ruled out of court.

To summarize: Geocentrism is the cosmology at which one arrives by way of cognitive sense perception, whereas Einsteinian acentrism corresponds to the way of knowing native to the physical sciences. There can be no conflict, no contradiction between the two: the respective worldviews simply correspond to different perspectives, different *darshanas*, as the Hindus would say. However, geocentrism is the higher of the two, even as the corporeal plane is ontologically higher than the physical.[14] Cognitive sense perception, moreover, having access to essence, is able in principle to transcend the corporeal plane: to pass, in the words of St. Paul, from "the things that are made" to the "invisible things of God"—and beyond even these, to "His eternal power and Godhead." In a word, human perception opens in principle to the metacosmic realms (from whence it has descended), whereas the modus operandi of physical science confines us to a relational and indeed subcorporeal domain. Galilean heliocentrism, finally, is a bastard notion which spuriously confounds the two ways of knowing. One might add that there is also a traditional or authentic heliocentrism, which must not be confused with the Galilean; and this is what will mainly concern us in the following chapter.

Notes

1. *The Copernican Revolution* (New York: MJF Books, 1985), 7.
2. One must remember that the smaller the orbit, the more rapid the motion.
3. It is safe to say that all serious advocates of geocentrism today are Tychonians.
4. A sketch of the proof may be found in Thomas Kuhn, op. cit., 201-206.
5. Ibid., 212.
6. Quoted in Dean Turner, *The Einstein Myth and the Ives Papers* (Greenwich: Devin Adair, 1979), 154.
7. See Chapter 6, 119, and Chapter 7, 133-134.
8. I have dealt with this question in "The Pitfall of Astrophysical Cosmology," *Sophia* 7(2), 2001, 5-27, reprinted here as Chapter 7.
9. *Physics and Philosophy* (New York: Harper & Row, 1962), 41.
10. *The Philosophy of Physical Science* (Cambridge University Press, 1949), 110.
11. Ibid., 32.
12. One is reminded of René Guénon's remarkable claim to the effect that those who cannot conceive of all events as occurring simultaneously are debarred from the least understanding of the metaphysical domains. This obviously places the metaphysical realm at the antipode of the physical, in which no two distinct events are actually simultaneous (although they may appear as such relative to a particular reference frame). It can be said that as one ascends the ontologic *scala naturae*, temporal dispersion is progressively reduced and eventually transcended.
13. See my article "From Schrödinger's Cat to Thomistic Ontology," *The Thomist* 63 (1999), 49-63, reprinted here as Chapter 2.
14. See my article "*Sophia Perennis* and Modern Science," in *The Philosophy of Seyyed Hossein Nasr* (La Salle, IL: Open Court, 2001), 469-485, reprinted here as Chapter 1.

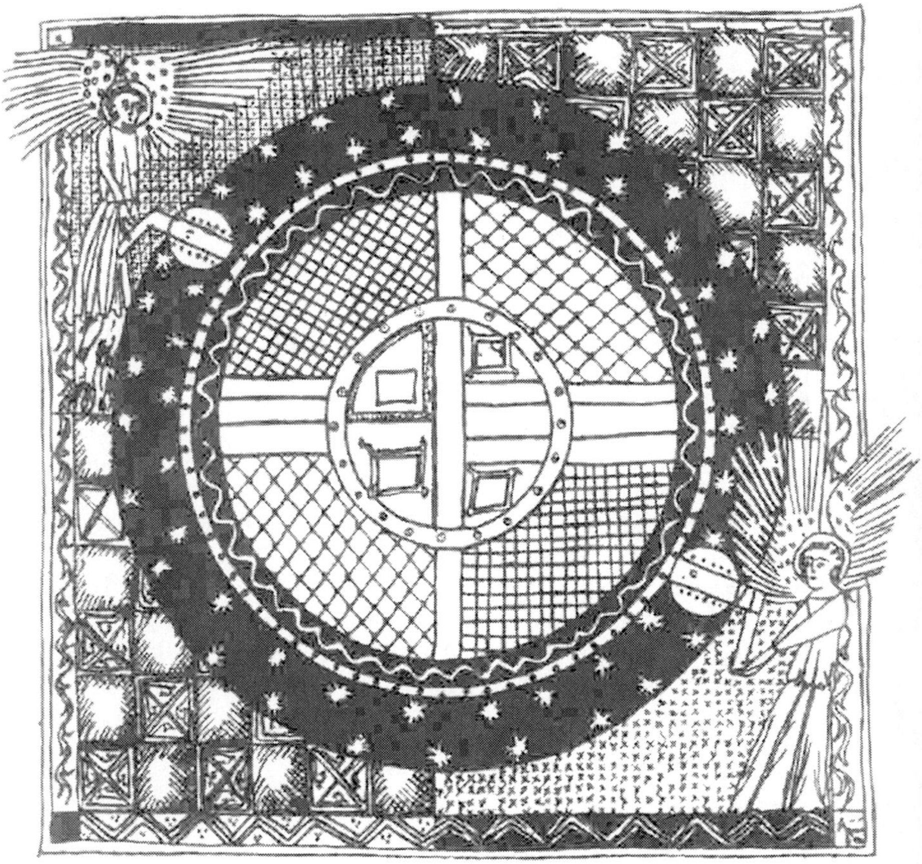

Angels turning the "first heaven". From a 14th c. manuscript.

Chapter IX

Esoterism and Cosmology: From Ptolemy to Dante and Cusanus

There are doctrinal conflicts which can only be resolved on an esoteric plane. In the present chapter, I propose to reflect upon one such conflict: the antithesis, namely, between a geocentric and a heliocentric worldview. It happens, however, that there is more than one geocentrism, even as there are several distinct kinds of heliocentrism. It is necessary, therefore, to sort out these various conceptions, which pertain to different levels and must not be confounded: only then can we grasp the crux of the problem.

In the first place it is needful, once again, to distinguish between two very different ways of knowing: the way of cognitive sense perception, which takes us into the corporeal domain, and the modus operandi of physical science, which gives access to what I term the physical universe. This said, it becomes apparent that the primary geocentrism—the geocentrism which is natural to mankind—is based upon the first way of knowing: looking up at the sky, one actually perceives the stars and planets circling the Earth, while the Earth itself is experienced as central and immobile. In regard to the second way of knowing, one generally takes it for granted that science has come down unequivocally on the side of heliocentrism. But as we have shown in the preceding chapter, it happens that contemporary physics does allow a geocentric hypothesis: the notion, namely, that the Earth does *not* move, does not indeed orbit around the Sun; according to Einsteinian relativity, no experiment can possibly prove otherwise. Admittedly, this is not much of a geocentrism; but so far as the scientific way of knowing is concerned, it is the most that can be said: physical geocentrism, let us call it, to distinguish the latter from the primary kind. To be sure, there is also a physical heliocentrism, which affirms that it is likewise admissible to consider the Sun to be at rest and the Earth to orbit around the Sun. On the level of physical theory, thus, there is no conflict between the two positions, which is to say that both derive support from the principles of relativity. As I have argued in Chapter 8, these principles, which appear to hold on the physical plane, are expressive of the fact that the notion of substance has no more place in fundamental physics: in a world in which only relations exist, I submit, Einsteinian relativity reigns supreme.

It should be noted that there is evidently no heliocentrism based upon cognitive sense perception. Nonetheless, apart from what I have termed physical heliocentrism, there is a celebrated heliocentrism championed by Galileo, which insists, supposedly on scientific grounds, that the Earth does move. One sees, however, that in claiming to have demonstrated the motion of the Earth, Galileo was in fact mistaken: his celebrated "Eppur Si Muove" remains to this day unproved. What I shall term Galilean heliocentrism, as I have noted before, turns out to be a bastard notion, a spurious hybrid, one can say, of the aforesaid two ways of knowing.

There is also, however, a third kind of heliocentrism, which might be termed traditional, iconic, and even perhaps esoteric; we will consider that heliocentrism in due course. But first it behooves us to reflect in some depth on the meaning and significance of the primary geocentrism.

It has been said that the geocentrist worldview is suited to the mentality of the so-called primitive man, someone who accepts the testimony of the senses uncritically, and is supposedly incapable of scientific thought. One maintains, moreover, that human perception is inherently unreliable and subject to manifold illusions, which need to be rectified through scientific means. Even scientists admit, of course, that sense perception does indeed constitute our one and only means of access to the external world; but one denies that it can *per se* bestow an authentic and accurate knowledge of things as they are. For that one needs to supplement the human faculties by scientific instruments, and avail oneself of the theories which underlie their use. The role of sense perception in the cognitive process is thus reduced ultimately to elementary acts, such as the reading of a pointer on a scale.

Oversimplified as this brief characterization of the science-oriented epistemology may be, it does serve to identify the contemporary scientistic denigration of sense perception as a serious and respectable way of knowing. To the scientistic mentality the modus operandi of science appears as the sole legitimate means for the acquisition of authentic knowledge; as Bertrand Russell once put it: "What science cannot tell us, mankind cannot know." But of course this is far from being the case! We need to understand from the outset that cognitive sense perception can give access to domains of reality beyond the range of scientific inquiry, and that in our daily life it does in fact give access to an authentic world which physical science as such cannot know. We need to remind ourselves that cognitive perception is neither a physiological nor indeed a psychological act, but is consummated in the intellect, the highest faculty within the human compound. So high, in fact, is that faculty, that according to Platonist philosophers it transcends the categories of space and time. Cognitive sense perception, thus, even in its humblest quotidian manifestations, proves to be something quite miraculous, something literally "not of this world." Moreover, in view of the fact that it constitutes our normal God-given means of knowing the external world, its scientistic denigration, I say, is not only fallacious, but impious as well. What actually limits the truth and the depth of human perception are not our faculties as such, but the use we make of them; and one should add that in this regard a collective decline appears to have been in progress since earliest times. It seems likely, moreover, that the scientistic denigration has itself had a debilitating effect upon our capacity to perceive, and has in fact accelerated our collective descent from the pristine state, a state in which, according to sacred tradition, man had the ability to penetrate "the things that are made" so as to apprehend "the invisible things of God" which they exemplify. The evolution of the scientistic outlook constitutes thus a late phase in that age-old descent which St. Paul has characterized as a "darkening of the heart." It is no doubt a fine line that separates true science from scientistic negation; yet we are told in no uncertain terms that those who cross that line are "without excuse." In words which appear to have lost none of their relevance, the Apostle describes the resultant condition of these perpetrators: "Professing themselves to be wise," he declares, "they became fools." (Rom. 1.20-22)

Having alluded to the collective decline which our powers of perception have suffered, it is to be noted that even in this diminished state we are yet able to behold a world that is truly sublime, and incomparably richer—and more real!—than the universe disclosed by the methods of physical science. To be sure, the scientific way of knowing has its validity and its corresponding ontological domain, as does the way of perception; but the latter, one is obliged to say, is the greater of the two. For it is by way of cognitive perception that we can know not merely the quantitative and material components of being, but can ascend to a knowledge of essences, and even, *Deo volente*, to a perception of "the invisible things of God."

Getting back to the question of geocentrism, it is to be noted that the worldview at which one arrives through sense perception is perforce geocentric. Now, in light of the preceding reflections, this fact, so far from constituting some kind of stigma, bestows in itself a certain legitimacy and indeed a certain primacy upon the geocentric *Weltanschauung*. One can say of the latter that it constitutes the normal human outlook, which as such cannot be illegitimate or void of truth. What we learn by way of our senses is that the Earth we stand upon reposes at the center of the universe, and that the Sun, Moon, planets and stars revolve around the Earth. It is true—as we have been told often enough—that the geocentrist outlook is suited to the understanding of simple and untutored minds; but it is equally true that this worldview is congenial to the understanding of sages and saints.

The traditional doctrine of geocentrism is based upon the conception of the Stellatum, the sphere of the stars, which rotates diurnally around the Earth. Between the Stellatum and the Earth there are the planets, the "wanderers," which differ visibly from the stars by the complexity of their apparent motions. What is of primary significance, however, is the underlying two-sphere architecture of the cosmos: the notion of an outermost sphere, comprised of stars, in perpetual revolution about the Earth, conceived as the innermost sphere. It is crucial to understand that the distinction between the two spheres, so far from being merely cosmographical, is primarily ontological, which is to say that the respective spheres represent two distinct ontologic domains, two worlds, if you will; and it is worth noting that to this day one speaks of "spheres" in a distinctly ontologic sense. It is likewise crucial to understand that the two worlds—the stellar and the terrestrial—define a hierarchic order: that the stellar world, namely, is "higher" than the terrestrial; and again I would point out that the adjectives "high" and "low" have to this day retained their hierarchic connotation. One sees thus that the two-sphere conception of the cosmos defines a dimension of verticality which is at once cosmographic, ontologic, and axiological. The immensity of spatial distance separating our Earth from the stellar sphere becomes thus indicative of the stupendous hiatus, both ontologic and axiological, separating the two domains. To be sure, the stellar world is not to be identified with the spiritual, which is metacosmic and invisible to mortal gaze; but yet, as the highest cosmic sphere, the stellar world reflects the spiritual to a pre-eminent degree.

According to ancient belief, there is an intimate connection between the stellar and the angelic realm, the realm of the so-called gods. The Earth, on the other hand, occupies the lowest position within the cosmic hierarchy, and this again is to be understood in a threefold sense.

These somewhat sparse indications may perhaps suffice to provide an initial glimpse of what geocentric cosmology is about. One sees that with his telescope and his polemics, Galileo had assaulted far more than a mere cosmography. It was not simply a question of whether the Earth does or does not move—whatever that might mean! Nor was it simply a question of whether the Galilean claim contradicts certain passages in Scripture, such as when the Good Book speaks of the Sun as "rising," or as "running its course." What stands at issue, clearly, is nothing less than an entire *Weltanschauung*. It is in fact the notion of cosmic hierarchy, of "verticality" in the traditional sense, that has come under attack. But let us note that this notion is intimately connected to the conception of spiritual ascent. One may object on the grounds that it is surely possible to "ascend" spiritually without flying up into the sky; but whereas the spiritual or metaphysical sense of verticality needs indeed to be distinguished from the cosmographic, it yet remains that the two are profoundly related. It is not mere imagination or pious poetry that Christ—and before Him, Enoch and Elias—"was taken up, and a cloud received him out of their sight." (Acts 1.9) The question remains, moreover, whether the two senses of verticality can in fact be separated on an existential plane, and whether the cosmographic sense may not indeed play a vital role in the spiritual life. One wonders whether an individual who thinks, *à la* Einstein, that "one coordinate system is as good as another," can in fact maintain a living belief in the possibility of spiritual ascent. What counts spiritually, as one knows, is what we believe with our entire being: inclusive, one is tempted to say, of the body itself, the corporeal component of our nature. Does not the First Commandment exhort us to love God "with all thine heart, with all thy soul, and *with all thy might*"? There can be little doubt that the ternary heart-soul-might corresponds to the Pauline *pneuma-psyche-soma*, which is to say that we are enjoined to love God not only with our spiritual and mental faculties, but with our corporeal being as well. Moreover, in line with this basic principle, the Church has decreed that the literal or "corporeal" sense of Scripture must not be denied,[1] must not be simply jettisoned, as contemporary theologians are wont to do. Authentic Christianity has always rejected angelism in any of its manifestations; if man is indeed a trichotomous being, his religious convictions and discipline need to be in a sense trichotomous as well. Getting back to the basic concept of verticality, it follows, then, that the cosmographic sense cannot be cast aside with impunity; and I would add that history appears to bear this out. It is surely not accidental that in the wake of the Copernican Revolution religious faith has visibly waned. In the more educated strata of society, at least, belief in the teachings of Christianity, to the extent that it has survived at all, has become strangely hollow, and conspicuously lacking in the force of existential conviction. There are notable exceptions, to be sure, but the overall trend is unmistakable; in a very real sense, Western man has forfeited his spiritual orientation. Having suffered the loss of cosmographic verticality, he finds himself in a flattened-out universe in which the concerns of authentic religion make little sense. Let it not be said that religion or spirituality have no need of a cosmology: nothing could be further from the truth. As Oskar Milosz has wisely observed: "Unless a man's concept of the physical universe accords with reality, his spiritual life will be crippled at its roots": yes, it is happening before our very eyes! Getting back to Galileo and his famous trial, one cannot but commend the Church for rallying to the defense of a position which in truth is its own.

It is vital to understand that geocentric cosmology is inherently an iconic doctrine. It pertains thus to the traditional sciences as distinguished from the modern, which are concerned with the material and thus non-iconic aspects of cosmic reality. As Seyyed Hossein Nasr explains:

> The modern sciences also know nature, but no longer as an icon. They are able to tell us about the size, weight and shape of the icon and even the composition of the various colors of paint used in painting it, but they can tell us nothing of its meaning in reference to a reality beyond itself.[2]

This is a very apt illustration, and a most enlightening one. A great deal of misunderstanding and confusion in the debate over geocentrism could have been avoided if the disputants on both sides had realized that the geocentrist claim is to be understood as an iconic truth, a truth which transcends the domain of the modern physical sciences. In reality geocentrism has to do with meaning, with cosmic symbolism, and thus with the mystery of essence. It is not a truth which can be defined, let alone demonstrated, on a positivistic plane.

Having characterized geocentrism as an iconic doctrine, it may be well to point out that what stands at issue is not a matter of symbolism in some psychological sense, but a matter, rather, of objective truth. Geocentrism is thus a *scientific* doctrine, one which pertains, as I have said before, to the province of the traditional sciences. As such it demands a certain ability to "see," to enter into a superior mode of vision, a mode that is able to discern the meaning of the icon as distinguished from mere "shapes and colors." The contemporary scientist, on the other hand, has been trained to fix his gaze precisely upon the outermost aspects of corporeal reality: is it any wonder that he misses the iconic sense? After considerable schooling one learns to reduce the icon to mere shape and color: reduce the universe, that is, to its material and quantitative components. And so it comes about that the true meaning of geocentrism generally escapes not only its scientific critics, but its contemporary scientific defenders as well.[3] The debate rages, when it does, over the outer husk.

Not only the reality, however, but the very conception of science in the traditional sense, has been virtually lost in the modern West. Even theologians, who *should* know better, have for the most part not a clue: if they had, they would not have busied themselves with the task of "demythologizing" sacred texts. Why this blindness? It is not a question of erudition, or even perhaps of "faith" in the religious sense; what is needed is a traditional ambience, something which in the West disappeared centuries ago. Nasr is no doubt profoundly right when he compares the traditional sciences to "jewels which glow in the presence of the light of a living sapiential tradition and become opaque once that light disappears."[4] We need to realize that this marvelous metaphor applies not only to various recondite disciplines, such as alchemy or astrology, but likewise to geocentrism, the meaning of which everyone presumes to understand. Given that cosmic realities are connected to their exemplars by way of essence, it follows that a worldview in which essence has been lost is one in which no traditional science—be it geocentrism or any other—can find recognition. Such a science may of course survive in its outer forms, even as the shapes and colors of an icon remain visible when its meaning has been lost. Geocentrism, in particular, may survive in its cosmographic dimension; thus reduced, however, to its external sense, it becomes in effect a superstition: a mere vestige of a forgotten worldview. In terms of Professor Nasr's metaphor, geocentrism has thus become "opaque."

Geocentric cosmology, whether conceived Ptolemaically or according to the Tychonian system,[5] affirms that the stars and the seven classical planets—Saturn, Jupiter, Mars, Sun, Venus, Mercury and Moon—are engaged in ceaseless revolution around the Earth, as if mounted on giant rotating spheres. In short, the heavens revolve while the Earth stands still: what is the significance of that? To the ancients it meant that the stars and planets are principles of motion in the terrestrial sphere. Even as the Sun gives rise to the alternation of day and night, and of the seasons, and the Moon gives rise to oceanic tides and other phenomena, so it is with the stars and the five remaining planets: such was the ancient belief. Astronomy and astrology were thus bound together as complementary aspects of a single science. One must not forget that Ptolemy has left us not only his *Almagest*—the most comprehensive and influential treatise on astronomy produced in antiquity—but also the *Tetrabiblos*, which deals with predictive astrology.

Given that the celestial spheres do indeed exert an influence upon the terrestrial world, how, let us ask, is that influence transmitted to the sublunar realm? At the hands of Aristotle this question received a rather physical answer: Having convinced himself on philosophical grounds that there can be no such thing as empty space, and persuaded that the celestial spheres are composed of an element termed the aether, Aristotle thought that each sphere exerts a kind of mechanical force upon the next, from the Stellatum down to the terrestrial. And since the latter sphere does not move, the result must be a mixing of the elements, and thus the production of internal motion and change: such, at least, is the apparent sense of the Aristotelian doctrine. It seems, however, that earlier conceptions of stellar influence had been far more theological than physical, if one may put it so; we must remember that preceding civilizations had populated the heavens with gods—or angels, as we prefer to say—who presumably disposed over more spiritual means of communicating their influence to the sublunar realm. But be this as it may, the celestial spheres were evidently conceived as "active" in relation to the terrestrial, which is to say that the worldview of these early civilizations was inherently astrological.

This basic feature of ancient cosmology has of course been abandoned in the wake of the Copernican Revolution. Copernicus himself tried hard to salvage as much as he could of the old cosmology; he was by no means a revolutionary or an iconoclast. Yet, by a kind of relentless logic, his astronomical innovation did precipitate the collapse of the ancient worldview: in the minds and imagination of those who, following Copernicus, came to espouse the heliocentric cosmography, astrology became a dead issue. For now the Earth itself revolves, and presumably acts upon other planets, even as these act upon the Earth. The new cosmology is visibly democratic: the traditional hierarchy, in which the Earth had been relegated to the lowest position, has been replaced by a planetary system in which the terrestrial globe enjoys more or less equal status with its companion planets. There is now no more "up" and "down," no more "east" and "west," "north" and "south," except of course in relation to a particular planet orbiting the Sun. Clearly, the very basis for an astrological outlook has disappeared.

In the new cosmology, the stars and classical planets no longer exert an influence upon the Earth; or to put it more accurately, no longer exert a "higher" influence. According to contemporary physics, there is an interaction via gravitational and electromagnetic forces; and certainly, in that sense, the Sun, Moon and stars still affect the Earth. But it is needless to point out that the action of forces or exchange of particles admitted by the physics of our day are nothing like the "influence of the celestial spheres" as conceived in ancient lore—which is of course precisely the reason why the very idea of astrology appears to us today as a primitive and indeed exploded superstition.

Iconic truth has to do with the relation of a cosmic to a metacosmic reality. However, since every cosmic entity is related to the metacosmic realms in multiple ways, it exemplifies a multiplicity of iconic truths. To read a cosmic icon, therefore, it is needful to make a choice; or better said, to engage in a particular perspective or point of view. What one beholds depends, so to speak, upon one's angle of vision; and as we change our point of vantage, the resultant perception may formally contradict the preceding cognition.

Having spoken of geocentrism as an iconic doctrine, I would like now to point out that heliocentrism, rightly understood, constitutes an iconic doctrine as well. The two seemingly rival contentions, thus, are both correct, which is to say that each embodies an iconic truth; it is the perspective, the point of view, that differs. More precisely, the two doctrines correspond to different *levels* of vision. The heliocentric position corresponds evidently to a more intellectual or internal kind of vision, inasmuch as it contradicts what might be termed the testimony of sense perception. Its iconic truth, moreover, derives from the fact that the Sun, as the representative of Deity, does by right occupy the center of the universe. As "the author not only of visibility in all visible things, but of generation and nourishment and growth"[6], it could not be conceived Ptolemaically as a mere planet, one among several that revolve about the Earth. Considering the overtly theophanic, one might almost say, "liturgical" outlook of the traditional heliocentric orientation, it is hardly surprising that heliocentrism has been especially associated with the Pythagorean and Platonist traditions, as opposed to the Aristotelian. Based on the report of Philolaus, the Pythagoreans espoused a non-geocentric cosmology in which the Earth revolves around a central fire, the so-called Altar of the Universe, which however was apparently not identified with the Sun. That identification came about later at the hands of the Neoplatonists, whose cosmology thus became overtly heliocentric. Later still, in the Renaissance movement championed by Marsiglio Ficino, the doctrine came alive again, but in a somewhat altered form; one might say that what Ficino instituted was indeed a religion, a kind of neo-paganism. Copernicus himself was profoundly influenced by this movement, as can be clearly seen from numerous passages in the *De Revolutionibus*. To cite but one example (from the tenth chapter of the First Book) which enables us to savor the spirit of those Renaissance times:

> In the middle of all sits the Sun enthroned. In this most beautiful temple, could we place this luminary in any better position from which he can illuminate the whole at once? He is rightly called the Lamp, the Mind, the Ruler of the Universe; Hermes Trismegistus names him the Visible God, Sophocles' Electra calls him the All-seeing. So the Sun sits as upon a royal throne ruling his children the planets which circle round him.

Yet despite these panegyrics, it appears that the light of iconic truth was fast fading. A kind of earth-bound literalism, hostile to the spirit of Platonic philosophy, was beginning to manifest itself, foreboding the advent of the modern age. Neither in Marsiglio Ficino nor in Copernicus do we encounter an authentic revival of Platonist doctrine, nor can it be said that the resultant heliocentrism conforms altogether to its traditional prototype: "rather was it comparable," writes Titus Burckhardt, "to the dangerous popularization of an esoteric truth."[7]

It behooves us to ponder this highly significant statement. Why should the truth of heliocentrism be "esoteric"? And why should its popularization be "dangerous"? We have already characterized the truth of authentic heliocentrism as "iconic"; are we perhaps to conclude that "iconic" and "esoteric" are one and the same? But by that token, authentic geocentrism would be "esoteric" as well. I propose to give at least a partial answer to these questions. Let it be noted, first of all, that there is a *prima facie* opposition, a kind of logical contradiction, between the geocentric and the heliocentric claims. It is to be noted, furthermore, that heliocentrism is based upon an intellective vision which replaces or supersedes the sensory. The crucial point, however, is that authentic heliocentrism does not deny that sensory truth, but accommodates it, rather, within an enlarged and perforce hierarchic vision of reality. Vivekananda has put it well when he said that "Man does not move from error to truth, but from truth to truth: from truth that is lower to truth that is higher." This toleration and indeed recognition of lower truth, I say, constitutes a mark of authentic esoterism. The higher truth is never destructive of the lower: quite to the contrary! A so-called esoterism, therefore, which undercuts the normal and in a sense God-given beliefs of mankind is perforce a false esoterism. Christ Himself has said: "I am not come to destroy, but to fulfill." And by way of further emphasis, He added: "For verily I say unto you, till heaven and earth pass, one jot or one tittle shall in no wise pass from the law, till all be fulfilled." (Matt. 5.17,18) To be sure, Christ is speaking presumably of the Mosaic law, and not of cosmology; yet even so I surmise that His words do also apply to the body of basic beliefs grounded in the Old Testament tradition, which certainly includes geocentrism. Till "heaven and earth pass," all these "lower truths" shall remain effective and in a way binding upon us: let no man cast them off before he has actually attained the higher—before "heaven and earth have passed away"—on pain of falling into what an Upanishad calls "a greater darkness."

Getting back to the *prima facie* contradiction between the geocentrist and the heliocentrist claims, I would like now to point out that this conflict cannot be resolved on the level of our ordinary "common sense" views concerning corporeal reality. Nor indeed can it be resolved on an Aristotelian basis, let alone a Cartesian. It needs, I surmise, to be resolved on the ground of a Platonist—or if you will, a Vedantist—metaphysics: no lesser realism, it appears, will suffice. And yes, that ground is indeed "esoteric," to say the least.

There can be little doubt, moreover, that this too is the ground upon which Dante conceived his monumental vision of what might be termed the integral cosmos. In a single poetic cosmography he combined, if you will, the geocentrist and the heliocentrist cosmologies; and it is highly significant that one passes from the former to the latter precisely at the threshold of the Empyrean, which thus represents the boundary, as it were, between the two "worlds." For indeed, as one crosses that boundary, the ascending spheres no longer expand, but now contract; in that supernal and indeed angelic realm, the hierarchic ordering of successive spheres is reversed: here to "ascend" means to approach the center, where stands the Altar of the Universe, the Throne of God. The Empyrean, thus—the outermost Ptolemaic sphere—marks the point of reversal, where "heaven and earth shall pass," which is also the point where "a new heaven and a new earth" shall come to be.[8] (Is. 65.17, Rev. 12.1)

The question arises whether the preeminence of authentic heliocentrism may not be reflected on the physical plane in some corresponding cosmographical preeminence. Does not the very principle of cosmic symbolism demand that the superior glory of the true heliocentric vision be mirrored somehow in the actual geometry of the planetary system? I submit that what Copernicus refers to as "a wonderful symmetry in the universe, and a definite relation of harmony in the motion and magnitude of the orbs, of a kind not possible to obtain in any other way," is none other than that reflection. Admittedly, the Copernican and the Tychonian systems prove to be mathematically equivalent,[9] which is to say that they predict the same apparent orbits; yet even so, the symmetries and harmony of which Copernicus speaks with justified ardor remain hidden in the Tychonian scheme, while they become resplendently manifest in the Copernican.

One has mixed feelings, therefore, concerning the contemporary defense of geocentrism. Christian believers do well in guarding a doctrine which proves to be basic to their faith; but the reductionist spirit of the times has forced the debate onto a cosmographic plane where the essential has already been lost, and where the defenders find themselves at a distinct disadvantage. As I have noted before, the principle of relativity has offered a certain protection to the beleaguered Tychonians; but at the same time it has rendered the geocentrist cause hopeless on physical grounds. Meanwhile the fact remains that a heliocentric coordinate system offers undeniable theoretical advantages precisely because it is adapted to the symmetries Copernicus had his eye upon: the very symmetries that bear witness to the heliocentric truth. The Tychonians may be right in claiming that they too can explain the observable facts, but one wonders at what cost in the form of *ad hoc* interventions.[10] There is something pathetic in the spectacle of these defenders, whom the opposing side does not deem worthy even of a response.

What necessarily baffles the exoterist mentality is what might be termed the multivalency of authentic revelation, be it scriptural or cosmic. Truth is hierarchical, and so Scripture and the cosmos itself need be in a sense hierarchical as well. No single perspective or level of understanding, no single "*darshana*," can do full justice to the integral truth: revelation itself informs us of this fact in various ways. Typically both Scripture and the cosmic revelation do so by way of "fissures," that is to say, by way of seeming incongruities which disturb and puzzle us, and hopefully spur us on to seek a higher level of truth. As Christ Himself intimated to His disciples on the eve before His passion: "I have yet many things to say unto you, but ye cannot bear them now." (John 16.12) Humility in the moral sense is not enough: we need also an intellectual and indeed theological humility. To preserve ourselves from falling into some arid dogmatism, we need ever to continue on our way: "from truth that is lower to truth that is higher." Dogmas, it seems, are meant for the *viator*, the spiritual traveler, not for the armchair theologian. It is not that dogmas of a sacred kind are simply provisional or limited in the ordinary sense, but rather that they harbor unsuspected truths. We need, as I have said, to continue on our way; as the author of Hebrews points out: "Strong meat belongeth to them that are full of age." (Heb. 5.14) Moreover, since truth derives ultimately from God, this progressive ascent constitutes indeed an *itinerarium mentis in Deum*, a "journey into God." But clearly, it is a journey in which the *viator* himself is progressively changed; in the words of St. Paul: "But we all with open face beholding as in a glass the glory of the Lord, are changed into the same image from glory to glory, as by the Spirit of the Lord." (2 Cor. 3.18)

As I have noted before, the higher truth of heliocentrism is reflected in the superior beauty or "symmetry" of the corresponding mathematical description; but one must remember that the "high truth" in question pertains to what may indeed be characterised as an esoteric level of vision. Reduced to a scientific theory in the contemporary sense—a mere cosmography—heliocentrism ranks in reality *below* its geocentric rival; as I have pointed out, the latter doctrine, limited though it be, corresponds to the testimony of human sense perception, and opens therefore upon vistas of truth which must remain forever unknown to the physical scientist as such. The problem with an "exoteric" geocentrism, on the other hand—a geocentrism that simply denies the heliocentric truth—is that it ultimately lacks a credible defense against a scientific heliocentrism: referents and epicycles, figuratively speaking, do not stand up well against the equations of Kepler and Newton. Even the most committed geocentrist can hardly fail to recognize a superior cogency in the heliocentric theory, and secretly sense that another truth must stand at issue, a truth which is not comprehended within the geocentric outlook. But alas, for the exoterist mentality that other truth becomes perforce hostile, perforce threatening to the integrity of the geocentric worldview. What by right should spur us on to seek a higher, more comprehensive level of understanding—what by right should be liberating—comes thus to be feared and rejected as a rank heresy.

The situation, however, is further complicated by the circumstance that heliocentrism has generally come to be identified with the Galilean doctrine, which *is* in fact a rank heresy. I have already argued that Galilean heliocentrism erodes the sense of verticality which supports and indeed enables the spiritual life: that it plunges us into a flattened and de-essentialized cosmos in which the claims of religion cease to be credible. I propose now to consider another ill effect of the Galilean heresy, which in a way is complementary to the aforesaid loss of verticality.

Every religion is perforce *homocentric* in its worldview. To put it in Christian terms: Man occupies a central position in the universe because he is made in the image and likeness of Him who is the absolute center of all that exists. Furthermore, man is central because, as the microcosm, he in a way contains within himself all that exists in the outer world, even as the center of a circle contains in a sense the full pencil of radii. Or again, man is central because he is the most precious among corporeal beings. In fact, Genesis teaches that God created the Earth as a habitat for man, and the Sun, Moon, and stars "for signs, and for seasons, and for days, and years." It is on account of man's centrality, moreover, that the Fall of Adam could affect the entire universe. Now, it is true that the centrality of which we speak is above all metaphysical, or mystical, as one might also say; yet even so, it is in the nature of things that this "essential" centrality should be reflected cosmographically. Does not the outer manifestation invariably mirror the inner or essential reality? To suppose that man can be metaphysically central while inhabiting a speck of matter occupying some nondescript position in some nondescript galaxy—that would surely be incongruous in the extreme. Once again: it would deny the very principle of cosmic symbolism, and thus the theophanic nature of cosmic reality. To be sure, it is possible, on an abstract philosophic plane, to affirm metaphysical centrality and cosmographic acentrality in the same breath; I doubt, however, that one can do so on an existential level, that is to say, in point of actual credence. To the extent that we truly believe the stipulated acentrality of the Earth, we are bound to relinquish the traditional claim of homocentrism: in reality, I say, these two articles of belief are mutually exclusive. One can, of course, pay lip-service to both, as a contemporary theologian might do; but actual belief—that is something else entirely.

The objection may be raised that it is indeed possible to espouse an acentric cosmology without detriment to the rightful claims of religion; and one might point to Nicholas of Cusa by way of substantiating that contention. True enough! One needs however to understand that the Cusan cosmology is profoundly Platonic, and corresponds, once again, to an authentically esoteric point of view. Its so-called acentrality is consequently worlds removed from the contemporary relativistic acentrality, and could be more accurately termed a "pancentrality." By the same token, moreover, the Cardinal does not simply deny the geocentrist claim, as does the Galilean astronomer: in reality he transcends the geocentrist contention, and in so doing, paradoxically, justifies and founds it "in spirit and in truth." "It is no less true," declares Nicholas of Cusa, "that the center of the world is within the Earth than that it is outside the Earth"; for indeed, "the Blessed God is also the center of the Earth, of all spheres, of all things in the world."[11] Here, in this terse and lucid statement worthy of a sanctified mind, we breathe the pure and invigorating air of a Christian esoterism. It is ever the way of authentic esoterism to "deny" only by affirming a higher truth, which contains but yet vastly exceeds the original claim.

It is true that the Earth enshrines the center of the universe; but so do the Sun, the Moon, and the myriad stars. Yet it is evidently the first of these recognitions that matters most to us so long as we are denizens of this terrestrial world. As I have noted before, we depend upon that recognition, that truth, for our orientation: our spiritual orientation no less than our physical.

What happens, now, when we ascend from a geocentric to an authentically heliocentric worldview: do we retain the original homocentrism? One may surmise that as we transcend the geocentric outlook, we likewise transcend the lesser theological conception of homocentrism, in accordance with the Pauline dictum: "I live, yet not I, but Christ liveth in me." (Gal. 2.20) The resultant and indeed higher homocentrism is in reality a Christocentrism; but again, that Christocentrism is not destructive of the earlier notion, the lesser truth—even as the Christ who "liveth in me" is not destructive of the "I" that "lives." It is once again a question of levels, of hierarchy. Meanwhile the intrinsic connection between geocentrism and the lesser homocentrism endures on the plane to which either notion applies, which is none other than the plane corresponding to our human condition. Let no one therefore deny either of these notions, either of these truths, "from below": the consequences of that denial cannot but be tragic in the extreme. Such a denial of either truth affects and indeed "poisons" every aspect of human culture, beginning with the life of religion, which it undermines.

It would be hard to overestimate the impact of the Copernican Revolution upon Western culture. Already in 1611, when the Revolution had barely begun, John Donne appears to have divined its larger significance: "And new philosophy calls all in doubt," he laments; "'Tis all in pieces, all coherence gone." No wonder the ecclesiastical guardians of the Roman Church were apprehensive as well, without perhaps realizing in full clarity what it is that ultimately stands at issue. Today, four centuries later, what lay concealed in that beginning has become clearly manifest, for all to see;

as Arthur Koestler has said, it is "as if a new race had arisen on this planet." Could this be the reason why St. Malachy, in his famous prophesies, has characterized the reign of Pope Paul V (1605-28) by alluding to the birth of "a perverse race"? One needs to recall that what is sometimes termed the first Galileo trial took place in the year 1616. What, then, could be the "perverse race" to which the saintly prophet refers? Given that Galileo is indeed "the father of modern science," one is compelled to answer that it is none other than the race of modern scientists, and by extension, the community of individuals imbued with the modern scientistic outlook. This, then, constitutes the fateful "birth" which took place during the pontificate of Paul V: no wonder St. Malachy has singled out the event! It was obviously not a question simply of planetary astronomy: what came to birth was a radically new outlook, a "new philosophy," as John Donne was quick to realize. From that point onwards, Western man began to look upon the universe with different eyes; and thus he found himself, quite literally, in a new world. Goethe, as always the realist, surely did not overstate the case when he declared that "probably not a single fact has had a deeper influence on the human spirit than the teaching of Copernicus." Only one should add that whereas Copernicus proposed the heliocentric hypothesis—the new mathematical model, if you will—it was Galileo who supplied the new philosophy.

As everyone knows, Galileo was formally tried in 1633 and forced to recant his Copernican convictions. The proposition that the Sun constitutes the immobile center of the universe was declared to be "formally heretical, because it is expressly contrary to the Holy Scriptures." And so the matter stood until 1822, when, under the reign of Pius VII, the Church commenced to soften its stand with regard to what it termed "the general opinion of modern astronomers." Thus began a process of accommodation with "the new race" which came to a head in 1979, when Pope John Paul II charged the Pontifical Academy of Sciences to re-open the Galileo case, and if need be, to reverse the verdict of 1633. Given the mentality which came to the fore in the wake of Vatican II, the outcome of that inquiry was never in doubt: Galileo was exonerated—some would say, "canonized"—following which Pope John Paul II in effect apologized to the world for wrongs committed by the Church. Could this be the reason, perhaps, why St. Malachy alludes to this Pope in the enigmatic words "*De Labore Solis*"? To be sure, the phrase, which traditionally refers to the movement of the Sun, does relate to Galileo, the man who denied that the Sun does move. Could it be, then, that St. Malachy, having previously signaled the birth of a "perverse race," is now alluding to the fact that some four hundred years later the Church has reversed its stand and relinquished its opposition to that "race," which is to say, to that new philosophy? Certainly St. Malachy's allusion can be interpreted in other ways as well; for example, "*De Labore Solis*" might be taken as a reference to the fact that this Pope, who has traveled far more extensively than any of his predecessors, has so many times "circled the globe" in his papal airliner (named, interestingly enough, "Galileo").

But be that as it may, the fact remains that the Church has now joined the rest of Western society in adopting a scientistic worldview; during the reign of Pope John Paul II, and with his sanction, a Copernican Revolution has finally taken place within the Church itself. Yet, to be precise, it is not the Church as such that has undergone change—that has "evolved," as the expression goes—but what has changed is simply the orientation of its human representatives: it is Rome, let us say, that has reversed its position. Humanly speaking, the ecclesiastic establishment may have opted for the only viable course: given the sophistication and prowess of contemporary science—given the "great signs and wonders" that could deceive even the elect—it may not indeed be feasible to stem the mounting tide of scientistic belief. Nonetheless one must insist, in light of our preceding analysis, that the contemporary cosmology, in any of its forms, is not in fact compatible with Christian doctrine. To the extent, therefore, that Rome has embraced a scientistic outlook, it has compromised the true teaching of the Church: this is the crux of the matter. Call it human failing, call it "political correctness," call it apostasy—the fact is that Rome has become "a house divided against itself."

Notes

1. As I have previously noted, in a ruling on "The Historical Character of the Earlier Chapters of Genesis," delivered in 1909, the Pontifical Biblical Commission denied the validity of "exegetical systems" which exclude the literal sense of Genesis. See Henry Denzinger, *The Sources of Catholic Dogma* (London: Herder, 1957), 2121-2128.
2. *The Philosophy of Seyyed Hossein Nasr* (La Salle, IL: Open Court, 2001), 487.
3. It may surprise some readers to learn that geocentrism still has scientific advocates. One of the best-known today is Gerardus Bouw, director of the Association for Biblical Astronomy, and editor of *Biblical Astronomer*, a journal dedicated to the scientific defense of geocentrism. See also his treatise *Geocentricity* (Cleveland: Association for Biblical Astronomy, 1992).
4. Op. cit., 488.
5. The so-called Tychonian system is named after its founder, Tycho Brahe, who proposed that Saturn, Jupiter, Mars, Venus and Mercury orbit around the Sun, while the Sun and Moon orbit around the Earth. It appears that virtually all contemporary defenders of geocentrist astronomy are Tychonians.
6. Plato, *Republic*, Book VI.
7. *Mirror of the Intellect* (State University of New York, 1987), 21.
8. On the subject of Dante's cosmology I refer to Titus Burckhardt, op. cit., 17-26, 82-98.
9. A sketch of the proof may be found in Thomas Kuhn, *The Copernican Revolution* (New York: MJF Books, 1985), 201-206.
10. So far as planetary orbits are concerned, Tychonian astronomy is equivalent to the Copernican. Ad hoc interventions, however, are required to account for stellar aberration and stellar parallax. See my article "The Status of Geocentrism," reprinted here as Chapter 8.
11. *On Learned Ignorance*, trans. Jasper Hopkins (Minneapolis: Banning, 1985), 115. A masterful discussion of the Cusan cosmology may be found in Jean Borella, *The Secret of the Christian Way* (State University of New York Press, 2001), Chapter 2.

Chapter X

Intelligent Design and Vertical Causality

Now, whatever lacks intelligence cannot move towards an end, unless it be directed by some being endowed with knowledge and intelligence, as the arrow is shot to its mark by the archer.

St. Thomas Aquinas,
Summa Theologiae I.2.3

From time immemorial, mankind has understood events or objects as the result of necessity, or of chance, or of design. These three basic categories of explanation appear to be native to the human mind, and as such they constitute what may be termed pre-philosophical notions. To be sure, there has been an ongoing effort on the part of philosophers to clarify these conceptions, and integrate them into a coherent account of causality. The simplest approach, perhaps, is to deny both chance and design, as the Greek atomists have done, and thus to suppose that all things occur by force of necessity, as Leucippus declared. Other schools, while still denying chance, have acknowledged design as a principle of causality not reducible to necessity; such was the case, for example, in the Stoic philosophy, which stipulated a kind of providential action or *pronoia* emanating from the World-Reason known as the Logos. Yet other schools of thought acknowledge both chance and necessity as irreducible principles, but deny design; and it is of interest to note that today many leading scientists espouse that position.

Among the three basic categories of causation, the most puzzling, perhaps, is the notion of chance. What confuses the issue, first of all, is the mistaken but commonly held belief that chance and necessity are irreconcilable: that it is a question simply of "either or." But clearly, to say that the toss of a coin yields heads or tails "by chance" is not to claim that the outcome has no cause, or is not in fact determined by its cause. Whether or not the toss of a coin is in some ultimate sense deterministic is a separate issue; what counts is that the event is, in any case, random or contingent in a suitably relative sense. Thus one finds that even in the heyday of classical physics, when the operations of Nature were deemed to be fully deterministic, statistical methods based upon the idea of chance could be successfully applied in various domains: for instance, in the kinetic theory of gases. The random distribution of gas molecules, and of their velocities, within a statistical ensemble, thus, does not contradict the supposition that the trajectory of each molecule is fully determined by a causal law. And I would point out in passing that this accords with Aristotle's idea that chance has to do with the coincidence of causally determined sequences of events, a scenario which occurs when two classical particles collide. Outside the domain of physical science, moreover, the notion of chance has always played an essential role. Courts of law, for example, distinguish regularly between accidental and non-accidental happenings, and insurance companies treat catastrophic events as a random variable with some given probability distribution. It is definitely meaningful, therefore, to speak of contingency and likelihood, regardless of whether Nature proves ultimately to be deterministic, or whether the outcome of every process is known beforehand to God. After all, we judge of things on the basis of *our* knowledge, and from our point of view; and as that knowledge, or that point of view, changes, so do our judgments relating to causality.

As concerns traditional doctrine, I contend that no orthodox school has been averse to the notion of chance. I have in fact argued on traditional grounds that there can be no such thing as a fully deterministic universe, and that contingency constitutes indeed a necessary complement of determination: that it represents in fact the *yin*-side of the coin.[1] It is needful that there be contingency as well as law. The concept of a clockwork universe, it turns out, is fatally flawed; and it is perhaps surprising that this discovery was made in recent times, not by poets or mystics, but by mathematical physicists, no less: by individuals who, of all people, are dedicated to the ideals of rigor and exactitude, and are thus predisposed towards the side of law. If such as these have reached the conclusion that contingency is necessary, after all, to the economy of Nature, this finding carries weight.

In light of these considerations one may combine necessity and chance into a single category under the title of natural causation, which then stands in contrast to design. In contrast, but not in opposition; for it is indeed an essential characteristic of traditional cosmology to admit both modes of causation: the divine, if you will, as well as the natural. It would thus be as contrary to the wisdom of tradition to maintain, on the one hand, that events and objects in the natural world are caused exclusively by divine action—as certain religious extremists have claimed—as it would be to maintain that the universe is governed simply by natural causes, as the philosophy of naturalism insists. Traditional cosmology, it can be said, acknowledges two seemingly opposed principles: the primacy of divine action, namely, and the efficacy of natural causes. I should remark, perhaps, that while I would not dispute Etienne Gilson's claim to the effect that the harmonization of these two principles has found its consummation in the Thomistic philosophy, I cannot accept his contention that Platonism, in particular, denies altogether the efficacy of natural causes, and thus affirms a radical extrinsicism.[2] To be sure, the philosophies of Plato and Aquinas represent different points of view, and it may be true that the efficacy of natural causes is affirmed more prominently in the teachings of the Angelic Doctor; but even so, I find Gilson's charge of "radical extrinsicism" to be misplaced. The point, perhaps, is that one can never understand a traditional philosophy which one does not approach with reverence. Suffice it to reiterate: *all* traditional cosmology, I contend, respects in the final count both the primacy of divine action *and* the efficacy of natural causes.

It has been likewise recognized, however, that manifest design cannot be attributed to natural causation. This is what St. Thomas Aquinas contends in his fifth proof for the existence of God; and let us note that the argument is validated, not by some abstract logic, but indeed on metaphysical ground, and thus on the basis of an intellectual perception. With the advent of modern times, however, the perennial "argument from design" has come under attack. First came deism, the "absentee landlord" philosophy which in effect exiled God from the universe; and this has led, by stages and degrees, to the full-fledged naturalism which came into vogue during the nineteenth century. As a dictionary of philosophy puts it: "Naturalism holds that the universe requires no supernatural cause and government, but is self-existent, self-operating, and self-directing; that the world-process is not teleological and anthropocentric, but purposeless and deterministic, except for possible tychistic events." Among the philosophical and theological movements which opposed this position, it was the school of British natural theology that centered its counterattack on the argument from design. "During the 17th and 18th centuries," we are told by Vergilius Ferm,

> there were attempts to set up a 'natural religion' to which men might easily give their assent and to offset the extravagant claims of the supernaturalists and their harsh charges against their doubters. The classical attempt to make out a case for the sweet reasonableness of divine purpose at work in the world was given by Paley in his *Natural Theology*, published in 1802.

Despite the fact, however, that this "natural religion" may have held its attraction for many an English gentleman, one finds that it fell woefully short of a tenable doctrine, due to the fact that it had in part assimilated the ambient naturalism which it wished to combat. In a word, British natural theology was a compromise solution, an eclectic doctrine that was bound to fall. And no one, it seems, knew better how to expose and capitalize on its weakness than Darwin himself: "I cannot persuade myself," he wrote, "that a beneficent and omnipotent God would have designedly created the Ichneumonidae [parasitic wasp] with the express intention of their feeding within the living bodies of Caterpillars."[3] True enough: there can be no answer to Darwin's objection, nor to the allied argument from dysfunction, without committing in some way to the doctrine of Original Sin and the resultant Fall—something which British natural theology, in its "sweet reasonableness," neglected to do. And so it came about, in the wake of Darwin's theory, that this "natural theology," which had enjoyed the approbation of an intellectual elite, succumbed eventually to the assault of naturalism.

We must not however lose sight of the fact that there is substance and indeed validity in Paley's conception of a "watchmaker God." If we walk through a field, as Paley invites us to do, and discover a watch lying on the ground, we may indeed conclude that this object is not the result of natural causes. It was not a natural law, nor a blind concatenation of accidental happenings, that artfully fashioned and assembled the parts of the watch to the end of keeping time. We are all perhaps familiar with a book entitled "The Blind Watchmaker," in which the Oxford zoologist Richard Dawkins proposes to refute Paley's claim; meanwhile, however, it turns out that the ancient argument has been recently revived on a scientific plane. The movement was sparked by Michael Behe, a molecular biologist, who introduced the concept of irreducible complexity, and argued convincingly that no natural process can give rise to structures which are in fact irreducibly complex. In *Darwin's Black Box*, published in 1996, Behe took his case before the general public and lucidly explained his position. His treatise, filled with captivating accounts of discoveries from the world of molecular biology, has become widely known, and has engendered serious debate in scientific circles. Behe's book, however, was only the beginning: the opening salvo, one might say, of a *scientific* counterattack, this time, against the prevailing naturalism. The decisive breakthrough was achieved two years later by a mathematician and philosopher named William Dembski, in a treatise entitled "The Design Inference." Dembski had asked himself the question whether design can perhaps be recognized by means of some signature, some criterion which can be defined in strictly mathematical terms. The resultant theory, it turns out, not only generalizes Behe's concept of irreducible complexity, but puts the question of a "design inference" on a mathematical—and hence rigorous—basis. What Dembski discovered is that a signature or criterion of design can indeed be given in terms of a probabilistic notion of *specified complexity*, or equivalently, in terms of an information-theoretic concept of *complex specified information* or CSI. The decisive result is a conservation theorem for CSI, which affirms in effect that CSI cannot be generated by any natural process, be it deterministic, random, or some combination of the two, as in so-called evolutionary algorithms. Thus was born a science termed design theory, also known as the theory of intelligent design or ID. The movement has of course drawn criticism from various segments of the scientific establishment, above all from the Darwinist contingent; but, in the final count, it is hard to argue with a mathematical theorem.

In the following section, I propose to present what will hopefully be a readable introduction to the theory of intelligent design. I should perhaps point out that I presuppose little in the way of mathematical background. Nonetheless, the "non-mathematical" reader may, if he so wishes, omit parts of this somewhat mathematical interlude, which is to say that the sequel is comprehensible on its own.

Let us begin with the concept of irreducible complexity: "By *irreducibly complex*," writes Behe, "I mean a single system composed of several well-matched parts that contribute to the basic function, wherein the removal of any one of the parts causes the system to effectively cease function."[4] The definition is evidently framed "with malice of forethought" to guarantee that no Darwinist process can ever give rise to an irreducibly complex structure; for as Darwin himself observed: "If it could be demonstrated that any complex organism existed, which could not possibly have been framed by numerous, successive, slight modifications, my theory would absolutely fail." To which however he added: "But I can find no such case."[5] The logic of Behe's argument appears to be impeccable: if a structure requires a number of "well-matched parts" before it can be functional, this precludes the possibility that "numerous, successive, slight modifications" could have been successively selected on the basis of function. The viability of Darwin's theory, therefore, does indeed hinge on the question whether one can "find such a case." Now, one of the most impressive and frequently cited examples of irreducible complexity is the so-called bacterial flagellum: a molecular device, whose function it is to propel the bacterium through its watery environment upwards along a nutritional gradient. The device consists of an acid-powered rotary engine—replete with a rotor, a stator, O-rings, bushings, and drive shaft—plus the actual flagellum, a kind of molecular rotary paddle. On account of the disorienting effect of Brownian motion, the flagellum must rotate at angular velocities on the order of 10,000 rpm, and must be able to reverse direction within one hundredth of a second. Moreover, to be functional, the device obviously requires auxiliary structures for detection and control, as well as for the production, storage, and distribution of the requisite fuel. What confronts us here, quite clearly, is a feat of nanotechnology that staggers the imagination; and needless to say, no one has yet proposed so much as the vaguest outline of a Darwinist scenario that might account for the production of these structures.

Yet, even so, Behe's argument remains incomplete. The conclusion that no Darwinist process could have produced the bacterial flagellum does appear to be inescapable; and yet the argument falls short of a rigorous proof. Now, the standard strategy, in the physical sciences, for proving that a closed system, operating under the action of natural causes, cannot attain a certain state, involves the use of an invariant satisfying a conservation law. It matters not whether the invariant is an energy, for example, which must remain constant, or a quantity, such as entropy, which cannot decrease (or increase); in either case, the conservation law rules out states that would violate that law. And let us note that this argument is perfectly rigorous, and does not require that we check out all possible scenarios which might conceivably bring the system into the disputed state. Getting back to Behe's contention, it is by means of this strategy, using CSI as his invariant, that Bill Dembski is able to refute the Darwinist claim in the case of structures such as the bacterial flagellum.

The basic idea of Dembski's theory is simple enough. Let us suppose that an archer is shooting arrows at a wall. To conclude that a given shot cannot be attributed to chance—in other words, to effect a design inference—one evidently needs to prescribe a target or bulls-eye which sufficiently reduces the likelihood of an accidental hit. What is essential is that the target can be specified without reference to the actual shot; it would not do, for example, to shoot the arrow first, and then paint a bulls-eye centered upon the point where the arrow hit. What stands at issue, however, has nothing to do with a temporal sequence of events: it does not in fact matter whether the target is given before or after the arrow is shot! What counts, as I have said, is that the target can be specified without reference to the shot in question. In Dembski's terminology, the target must be "detachable" in an appropriate sense. Consider a scenario in which the keys of a typewriter are struck in succession. If the resultant sequence of characters spells out, let us say, a series of grammatical and coherent English sentences, we conclude that this event cannot be ascribed to chance. An exceedingly unlikely and indeed "detachable" target has been struck, which however was specified *after* the event. In general, the specification of a target requires both knowledge and intelligence; one might mention the example of cryptanalysis, in which specification is achieved through the discovery of a code. What at first appeared to be a random sequence of characters proves thus to be the result of intelligent agency. The fact is that it takes intelligence to detect intelligent design.

I would like to emphasize that it is impossible to rule out the hypothesis of chance simply on the basis of low probability. If a sequence of 1's and 0's is generated by tossing a fair coin 1000 times, the possibility that the resultant bit string will contain not a single 0, let us say, can indeed be ruled out. Yet, if one does actually toss a coin 1000 times, one produces a bit string having exactly the same probability as the first: 1 in 2 to the power 1000, to be precise. Why, then, can the first sequence (the one containing no 0's) be ruled out, while the second can indeed occur? The reason is that the first conforms to a pattern or rule which can be defined independently; it is a question, once again, of a "detachable" target which itself has low probability. In the case of the first sequence, the prescription "no 0's" itself defines such a target: the subset, namely, containing the given bit string and no other. But this is precisely what can *not* be done in the case of the second bit string (the one produced by tossing a coin 1000 times): it is virtually certain in that case that no detachable target of low probability has been hit. It is possible, of course, to come up with a description by reading off the sequence itself; but that description or pattern (if such it may be called) will turn out not to be detachable. To read a description off the event is like painting a bulls-eye around the point where an arrow has struck: such a description, of course, proves nothing. The discovery of a *detachable* pattern of sufficiently low probability, on the other hand, proves a great deal: it proves in fact that the event in question cannot be attributed to chance. What rules out chance, thus, is not low probability alone, but low probability in conjunction with a detachable target: this winning combination is what Dembski terms the complex specification criterion.

The general mathematical structure within which design theory operates is as follows: One is given a reference class of possibilities Ω, together with a probability measure P which assigns to each (measurable) subset of Ω a real number between 0 and 1. Given an elementary event E (represented by a point in Ω), a specification of E is a subset T of Ω containing E, which is "detachable" in a sense to be defined. The pair (T, E) is then said to constitute a specified event. It is to be noted that a specified event has two components: a conceptual component T, one can say, and a physical component E. It constitutes a twofold entity, thus, a thing that combines, so to speak, two worlds. And therein, let me add, lies the power and indeed the genius of Dembski's theory: where others have dealt with events, Dembski deals with specified events, a categorically different kind of thing. Let us suppose, now, that one is able to associate an invariant I with each specified event, which cannot, say, increase under the operation of natural causes; as we have previously noted, such a conservation law could validate a theory of design. To obtain a suitable invariant, Dembski replaces the probability measure P by a corresponding information measure I, defined by the equation

$$I = -\log_2 P .$$

According to this formula, the information contained in a bit string of length n is just n. In general, for an event A in Ω, I(A) represents by definition the information content of A as measured in bits. It is to be noted that P and I are inverse measures: the smaller P, the larger I will be; indeed, as P tends to zero, I goes to infinity. In mathematical parlance, I is thus a measure of complexity. Dembski goes on to define the information content of a specified event (T, E) to be I(T): what counts is the conceptual component of the specified event. Finally, the information contained in a specified event is what Dembski terms specified information, and this is what he takes initially as his invariant.

It turns out, first of all, that specified information is strictly conserved under the action of a deterministic process. As one would expect, the proof hinges upon the fact that a deterministic process can be represented by a function that has the totality of initial states as its domain, and the resultant states at a later time t for its range. Let us suppose, then, that Ω is a reference class of possibilities, and that (T, E) is a specified event in Ω. If E is causally determined by an event E_0, there exists then a reference space Ω_0 containing E_0, and a function f from Ω_0 to Ω, such that $f(E_0) = E$. It is to be noted that subsets of Ω "pull back" under f, which is to say that f induces a function f^{-1} from the powerset of Ω to the powerset of Ω_0. One may therefore define a subset T_0 of Ω_0 as the inverse image of T; and as might be expected, it turns out that (T_0, E_0) constitutes again a specified event. Let it be noted, further, that a probability measure P_0 on Ω_0 induces a probability measure $P = P_0 f^{-1}$ on Ω, such that $P_0(T_0) = P(T)$. It follows that the specified events (T_0, E_0) and (T, E) carry precisely the same amount of specified information.

Surprisingly, this conclusion could have been foreseen without recourse to mathematical analysis; as Dembski has put it: "What laws cannot do is produce contingency; and without contingency they cannot generate information . . ."[6] The point was made four decades earlier by Leon Brillouin, when he wrote that "a machine does not create any new information, but it performs a very valuable transformation of known information"; and it appears that as far back as 1836, the poet and amateur scientist Edgar Allen Poe had said much the same.[7]

But what about random processes? As we have seen, a random process can generate arbitrarily large amounts of information (due to the fact that events of arbitrarily small probability can indeed occur), and it can even generate small amounts of specified information; what it cannot do, according to the complex specification criterion, is to generate specified information in large amounts: that is the crucial point. The proverbial monkey pounding on a typewriter can perhaps produce a few bits worth of English prose, but not the text of Hamlet. There must be a cutoff, even though its exact location cannot be ascertained. It can thus be stated that there exists a universal complexity bound or UCB, beyond which specified information cannot be increased by a random process taking place within the bounds of physical space and time. Dembski, for good reason, takes his UCB to be 500 bits, which is "playing it safe": no monkey could remotely do that well—not in a billion years! What no random process can generate turns out to be *complex* specified information or CSI: specified information, namely, in excess of the UCB.

Having thus disposed of deterministic as well as random processes, it remains to consider an arbitrary combination of the two. Now, any such combination can be modeled by a so-called stochastic process. In place of a function f(x) of a single variable, one has now a function $f(x, \omega)$ of two variables, in which ω represents the random component of the process. The trick is to break the problem into two parts: one first permits ω to "occur," which transforms the original function f into a function f_ω of a single variable x, in which ω serves as a fixed parameter. That function, however, can then be handled as in the deterministic case. One is able by this means to conclude that the total process cannot increase specified information by more than the UCB.[8]

What is the significance of this conservation law? Clearly, it shows that CSI cannot be generated by any natural process. In particular, the vast amounts of CSI existing within the DNA of a living cell could not have been produced through the Darwinist scenario of random variations acted upon by natural selection. Neo-Darwinists such as Manfred Eigen, having recognized the origin of information as the major problem of contemporary biology, have been searching for an evolutionary algorithm that can do the job, without realizing that this possibility can in fact be ruled out on theoretical grounds. Every evolutionary algorithm, no matter how ingeniously conceived, is a stochastic process, and as such it falls under Dembski's interdict. Only two possibilities remain: either the universe must have been replete with CSI from the first moment of its existence—a supposition which hardly accords with the hypothesis of a "big bang" origin—or else it cannot be conceived as a closed system operating under the action of natural causes.

It appears that CSI, and more generally, information in the mathematical sense, has a certain ontological status; it was Norbert Wiener, I believe, who first pointed out that, in addition to mass and energy, the universe comprises "information" as a basic ingredient. It is in any case to be noted that the course of science, from about 1900 onwards, has tended towards that recognition. The trend began with Boltzmann's statistical mechanics, which demonstrated that the notions of contingency and probability play a vital role in the economy of Nature. Contingency and probability, however, add up to information, as we have seen; and I might mention that the statistical definition of entropy, as given by Boltzmann, is actually formulated in information-theoretic terms. It was by way of Boltzmann's statistical approach to blackbody radiation, in particular, that Max Planck was able to discover the so-called quantum of action, nowadays known as Planck's constant, a discovery which inaugurated the quantum era.[9] With the advent of quantum mechanics, moreover, it became apparent that probabilities and information are not only useful conceptions, but prove indeed to be necessary on the most basic level of physical theory. It is true that Einstein, for one, refused to accept this conclusion; yet every argument which he advanced to counter the quantum-mechanical indeterminism proved ultimately to be ineffectual. More often than not, it actually contributed insights which only served in the end to bolster the quantum-mechanical position, as was the case with the well-known Einstein-Podolsky-Rosen thought experiment, which was later carried out, only to confirm the quantum-mechanical predictions. Moreover, when David Bohm, after prolonged discussions with Einstein, did finally succeed in constructing what appeared formally to be a deterministic quantum theory, he did so at the cost of introducing what he termed "active information" as a basic principle; but as Dembski was quick to point out, Bohm's "active information" is but a special case of CSI. It appears that Bohm was able to dispense with indeterminacy on the level of particles only by admitting contingency in the form of CSI; one way or another, it seems, contingency—and thus information—is bound to enter the picture. Meanwhile CSI has made a second appearance as a foundational concept of physics, this time in the form of the so-called Fisher information from which Roy Frieden claims to derive all the basic laws.[10] From another direction entirely, recent advances in communication and computer technology have given rise to a number of mathematical sciences in which information, normally in the form of CSI, plays a leading role. Information theory itself, as a mathematical discipline, was inaugurated in 1948 by Claude Shannon, an electrical engineer concerned with communication problems. But unquestionably the most significant encounter with information has taken place in the domain of molecular biology, which has brought to light what may be termed the primacy of CSI in the biosphere. The fact that vast quantities of specified information, recorded in a four-letter alphabet, reside within every living cell, and that each species derives, as it were, from a text known as its genetic code, suggests that life has indeed an informational basis.

Given the crucial role of CSI in both physics and biology, it behooves us now to reflect further upon that notion, beginning with the mathematical concept of information as such. The danger, when it comes to the latter concept, is that we are prone to read far more into the term than it is meant to signify: the word "information," after all, has obviously been in use for a very long time before Shannon gave it a technical sense. That sense is in fact quite bare: it boils down to the actualization of a contingency in a mathematical space Ω of viable possibilities, endowed with a probability measure P. If I flip a coin n times, I have produced information: n bits worth, to be exact. And even now, as I am striking the keys of my typewriter, I am producing Shannon information. I am also, however, generating *semantic* information, which is something else entirely, something which no mathematical theory can encompass, for the obvious reason that semantic information is not a quantitative thing. There is an ontological discrepancy, thus, between semantic and Shannon information, not unlike the ontological hiatus between the corporeal and the physical domains; and as in the case of a corporeal object X and its associated physical object SX, it does happen that every item of semantic information is associated with a corresponding item of Shannon information, which serves, so to speak, as its material base. The latter is what remains, one can say, after all meaning, all non-quantitative content, has been cast out or "bracketed." This accords, once again, with René Guénon's point that "quantity itself, to which they [the moderns] strive to reduce everything, when considered from their own special point of view, is no more than the 'residue' of an existence emptied of everything that constitutes its essence."[11]

Having thus distinguished between semantic and Shannon information, I would like to point out that the semantic component constitutes an example of specification, an instance of a "detachable target."[12] To be sure, the example of semantic information is highly special, which is to say that specification can arise in a thousand other ways. Think of a bit string in which 1's and 0's alternate, or in which they represent a sequence of prime numbers in binary notation; or again, think of a bit string of length n which is "algorithmically compressible" in the sense that it can be generated by a computer algorithm of "length" less than n (a notion which can indeed be defined): all these are examples of specification. However, despite its immensely special nature, I will note in passing that semantic specification enjoys a symbolic primacy in the natural domain. If it be the case that God "spoke" the world into being, as Scripture declares, such design as it carries must derive ultimately from a divine idea or *logos*, which may by analogy be termed a "word." And I would add that nowhere in the natural world is the linguistic character of specification more clearly in evidence than in the genetic code of an organism, which constitutes a text, as I have noted before, recorded in a four-letter alphabet. The genetic code, then, is a *written* text, imprinted on DNA; yet one may conclude on theological ground that this written text derives indeed from a spoken word, the kind to which Christ alludes when He testifies that the words He speaks "are spirit and life." (John 6.63) No other scientific finding, I believe, has been as profoundly reflective of theological truth as the discovery of what may be termed the informational basis of life.

Following these rather cursory considerations relating to the concept of CSI, I propose to reflect in some depth upon the nature of causality. From the outset we have distinguished between necessity, chance, and design, and have combined the first two modes under the heading of natural causation. On the strength of Dembski's theorem we may now conclude that CSI, wheresoever it may be found, must be attributed to an alternative mode of causation answering to the notion of design. I now contend that this alternative mode is none other than what I have elsewhere termed vertical causation, a mode characterized by its atemporality.[13] Inasmuch as vertical causation acts "above time," or instantaneously, as one can also say, it differs fundamentally from natural causation, which is inherently temporal, and could therefore be characterized (in metaphysical parlance) as "horizontal." One arrives thus at a dichotomy which needs now to be carefully examined and clarified.

The prime example of vertical causation is unquestionably the creative act of God; for as St. Augustine says: "Beyond all doubt, God created the world, not in time, but with time." One needs however to realize that this creative act extends in a sense to God's providence, the first effect of which is what theologians term "conservation." As Gilson points out, this effect "is, in some way, but the continuance of the creative act." We need not attempt to classify God's action upon the world; suffice it to say that God acts ever above time, and thus "vertically." That vertical causation, moreover, is the cause, not only of time, but also of the actions and processes that transpire in time. Yet these actions and processes have an efficacy of their own: such is the miracle of God's creation. God is intimately present, not only in the substance of all beings, but in their operations as well; and yet, as Gilson has beautifully said, "the intimacy of the assistance He gives leaves their efficacy intact." It follows that God's vertical causality is complemented by a causality which operates in time; and this is what we have termed horizontal causation: it is the kind, obviously, with which science is concerned. A word of caution, however, needs to be interposed: To say that every act of natural causality is horizontal is not to imply that every act of horizontal causality is natural. There may conceivably be temporal processes which are neither deterministic, nor random, nor yet stochastic, a fact which implies that the concept of horizontal causation is wider than that of natural causes. We shall have occasion to return to this point presently.

Having distinguished between a vertical causality, which is proper to God, and a horizontal causality which operates by way of a temporal sequence, we need to ask ourselves whether this dichotomy amounts simply to the traditional distinction between primary and secondary causation. The answer, clearly, is that it does not; for it happens that there exist second or created causes which likewise act above time, and thus vertically. What stands at issue is a higher degree of participation in the divine causality, one which preeminently reflects the action of God Himself. There are two prime examples of this higher mode of secondary causation: the causality, namely, emanating from the angelic realm, and in second place, action derived from human intelligence. Could it be that vertical causation is in fact the hallmark of intelligence? This appears indeed to be the case: to act "above time" is apparently the prerogative of an intellectual nature, a being endowed with intellect and free will. Vertical causation, thus, is none other than intelligent causation, whereas natural causes may be characterized by comparison as "blind."

This brings us at last to a question which has been lurking in the background from the start: the matter, namely, of human art, in the widest sense of that term. To the modern mind, at least, the most obvious and incontrovertible instances of design are those resulting, not from an act of God, but from the action of a human artisan. Not even the most committed Darwinist would deny the presence of design in the case of Paley's watch; but whereas he accepts the notion of design in the sphere of human artifacts, he considers it "naively anthropomorphic" to extend this notion to the biological domain. We are in fact surrounded on all sides by CSI deriving from intelligent human action, and long before Dembski appeared on the scene, it was clear to everyone that this CSI is not in fact the result of a natural occurrence, but was put there by an intelligent agent. Who, for example, when he comes upon a collection of stones on a hillside spelling out the word "Welcome," would imagine that these stones were deposited by a flood or an avalanche? In a thousand ways, all of us have been engaged, since early childhood, in the business of inferring design; and whether we know it or not, these inferences are invariably based upon specification. Dembski's theory, thus, applies in the first place to the domain of human art.

Let us then consider the production of artifacts, from primitive crafts to modern industry. Is it not obvious, one might say, that the artifact is invariably produced by means of a temporal process? In a sense this is of course true. I do not deny the necessity of temporal process; what I deny is its sufficiency. My contention is twofold: first, that the critical factor—the *sine qua non* of human art—is an act of intelligence; and secondly, that such an act is not reducible to a temporal process. Few, I suppose, would object to the first claim; it is the second that troubles us. The difficulty stems from the fact that we tend to temporalize the act of intelligence by identifying cognition with thought. We take it for granted that cognition occurs *within* thought—within a psychosomatic and temporal process—whereas in fact thought is only a means, a movement, if you will, in quest of cognition. To put it in traditional terms: cognition is an intellective as opposed to a psychosomatic act. As Aquinas observes: "The activity of the body has nothing in common with the activity of the intellect."[14] And this in itself suggests strongly that intellectual activity does not take place "in time." The crucial point, however, is that intellectual activity *cannot* take place in time, for the simple reason that temporal dispersion is incompatible with cognition: we cannot know "bit by bit," because to know is necessarily to know *one* thing. This conclusion cannot be obviated, moreover, by adducing memory as a means of presentifying the past; the fact remains that the cognitive act must be "instantaneous," and hence supra-temporal: for indeed, the moment or instant is not a part of time. The intellect, therefore, whether conceived as a "third principle," or (Thomistically) as a power of the soul, must be inherently supra-temporal as well. Getting back to the production of artifacts, the following has now become clear: If intellectual agency is indeed a *sine qua non* of human making, it is *ipso facto* impossible to reduce the production of the artifact to a temporal process. It is true, of course, that mechanized manufacture *is* a temporal process; but one must not forget that the machinery involved in this process carries design, which is transmitted to the resultant product. As Dembski's analysis shows, a deterministic process, and thus a function, may indeed transmit CSI, but cannot produce it (not even to the extent of a single bit!). A machine-made artifact, thus, no less than one produced by a human artisan, presupposes an act of vertical causation.

Human making is allied to God's creative act by virtue of the fact that it likewise entails an atemporal mode of causation. It can be said that the human artist imitates the divine, and "participates" to some degree in God's creative agency. "Art imitates Nature [in the sense of *natura naturans*] in her manner of operation," says Aquinas; and elsewhere he specifies that the human artist works "through a word conceived in his intellect," and thus in imitation of the Holy Trinity.[15] It is no small thing, therefore, that transpires in even the humblest instance of human making, which is indeed worlds removed from a mere temporal process. No wonder this difference can be detected in the artifact by way of a distinctive signature indicative of design.

A further clarification needs to be made. Having characterized vertical causation by the fact that it is atemporal in its mode of operation, we must bear in mind that it may nonetheless be temporal in its effect. A violinist, for example, does indeed act "above time" on the plane of intellect, and yet the music he plays is produced by a movement of his bow. Once again a temporal and thus horizontal process enters the picture; but clearly, it is not a natural process: it cannot be, since it derives from an intellectual act.[16] One needs therefore to distinguish (as I have intimated previously) between two kinds of temporal process or horizontal causation: the kind that derives from natural causes, and the kind that springs from intelligent agency. It is the violinist, acting as an intelligent agent, who first apprehends the music—Dembski's "detachable pattern"—on the plane of intellect, and then, by an act of his free will, conveys that pattern to the world of sense by way of a temporal process, an action of horizontal causality.[17]

A few words on the subject of "free will" are called for at this point. One sees from the example of the human artisan that intelligent action is not—and cannot be—the result of a natural process. The cause of such action cannot therefore be identified on the natural plane, that is to say, in the external world, and may consequently be characterized by default as "internal." But is that not indeed what we normally mean when we speak of "free will"? Now, if by "freedom of the will" we understand an exemption from external causality, then the preceding reflections do in fact establish that freedom in the context of human art. The decisive fact, however, is that intelligent action is "free" by virtue of an intimate participation in the freedom of God Himself. And this divine freedom, to be sure, is infinitely more than a mere exemption from external constraint; after all, it is by virtue of this very freedom that God created the world, and thus "external constraint" itself. Freedom, therefore, has primarily a positive connotation: it has to do with creativity, with the expression of truth and beauty, and also with "play," with what Hindu tradition terms *lila*.

One final point needs to be made: nothing obliges us to suppose that a temporal process which is *not* productive of design, or of CSI, can *ipso facto* be attributed to natural causation. A case in point is given by the quantum-mechanical phenomenon of state vector collapse: the radioactive disintegration of a radium atom, for instance, cannot in fact be accounted for in terms of natural causation by virtue of its irreducible discontinuity. The ancient dictum "*Natura non facit saltum*," I claim, holds to this day; only it needs to be understood that the *natura* in question is *natura naturata* as distinguished from *natura naturans*: the "natured" as opposed to the "naturing." I have argued elsewhere that *natura naturata* acts invariably by way of a *continuous* temporal process, in contrast to *natura naturans*, which acts "above time" and thus by vertical causation.[18] The action of *natura naturans* is therefore inherently instantaneous, and it is this intrinsic instantaneity, I contend, that is reflected in instances of irreducible discontinuity. The reason why state vector collapse has mystified physicists, it turns out, lies in the fact that the phenomenon cannot be attributed to natural causes. What disturbs the Einsteinians, it appears, is not merely the breakdown of determinism, but indeed the collapse of natural causation: not only does God "play dice," but what is worse, He does so "instantaneously"!

The fact that the DNA in a living cell carries "tons of CSI" implies that Dembski's theorem has disqualified the claims of Darwinism. It has dashed the neo-Darwinist hope of finding "an algorithm, a natural law that leads to the origin of information," as Manfred Eigen has put it.[19] What stands at the heart of a living organism is CSI, and no natural law, no algorithm, no stochastic process can produce that CSI. I will leave out of account the question as to how long it may take the scientific community at large to accept this fact and draw the consequences; certainly, if we admit what Thomas Kuhn has to say on the subject of "scientific revolutions," this will not happen overnight. What in any case I find to be of far greater interest than the rise and fall of the Darwinist paradigm is the fact that Dembski's theory poses a fatal threat, not just to Darwinism, but to the cause of authentic religion, surprising as this may seem. To be sure, theologians for the most part are jubilant to learn that God is not superfluous after all; yet what they fail to realize, almost to a man, is that the design movement threatens to plunge us into a heresy worse than Darwinism. The problem is this: Whereas design theory has indeed disqualified the Darwinian mechanism, it has in no wise discredited the Darwinist concept of common descent, which thus remains entrenched as a scientistic dogma. But clearly, the hypothesis of a common descent which cannot be accounted for in terms of the Darwinian mechanism—nor, for that matter, in terms of natural causation as such—is tantamount to the tenet of theistic evolution. It is almost inevitable, therefore, that Dembski's discovery will be generally perceived as a scientific vindication of that tenet, a doctrine which has already swept the theological world and penetrated even into the Vatican. Thus, if theistic evolutionism was the rage long before Dembski proffered his scientific insight, just think what its status will become in the wake of his monumental discovery! Here indeed is a doctrine to "deceive even the elect."[20]

The problem with the notion of common descent is that it obviates metaphysics in a domain that is incurably metaphysical. Common descent, to be sure, if there be such a thing, is something that transpires in space and time: it is something we can picture, something that answers to the demands of our ordinary understanding. Therein lies its appeal, and therein too lies its impossibility: for it happens, as every traditional school has recognized, that first origins cannot in fact be situated in space and time. There exists, for instance, a Patristic doctrine concerning first biological origins—the doctrine of *ratione seminales* elaborated by St. Augustine in *De Genesis ad Litteram*—but the teaching is irremediably metaphysical: it alludes to a vertical descent, a progression from the metacosmic Center to the cosmic periphery. But this means that the scientist is in fact "coming in" near the end of the story: confined, by the *modus operandi* of his approach, to an exclusively horizontal perspective, he misses the vertical descent which ontologically precedes all manifestation on the spatio-temporal plane. Thus, in keeping with this horizontal perspective, the hypothesis of common descent proposes to resolve the mystery of first biological origins within the spatio-temporal domain: at the felly of the cosmic wheel, where indeed first origins can never occur. Moreover, to bring God into the picture, as the theistic evolutionists have done, does not alter this fact, this principial impossibility: it only compounds bad science with bad theology.

One forgets that authentic evolution is indeed an unfolding, as we learn from the Latin verb *evolvere* (*e* + *volvere*, "to roll out"); it is thus an outbound kind of movement. Where there is an outside, however, there must also be an inside, an interior; and let me hasten to add that we must not psychologize this "inside": the *bona fide* interior of an organism is not a matter of "consciousness," but constitutes the ground from which every component of the organism, including consciousness itself, is derived. The integral organism—like the integral cosmos—may thus be conceived in terms of a symbolic circle, whose center represents its "innermost" point, the true *ratio seminale*,[21] of which the visible creature, situated in space and time, is but the outward manifestation or "unfolding." It is moreover of interest to note that the Greek equivalent of the term *ratio seminale* is *logos spermatikos*: the "seed word," if you will, whose marred reflection, as I have suggested previously, can indeed be discerned in the genetic code. The "origin of information," thus, which neo-Darwinists are seeking in an evolutionary algorithm, is actually to be found in the *logos spermatikos* that came into being in the single instant of creation, when "God created the heaven and the earth." From that point of origin there began a "vertical descent," an evolution in the true sense of the term, in which however the essence of the organism, its *ratio seminale*, remains unchanged. Here, then, is the crux of the matter: to comprehend this metaphysical truth—even as "through a glass, darkly"—is to perceive at once the fallacy, not just of Darwinism, but of theistic evolutionism as well. But the latter doctrine, as I have indicated before, is the worse of the two: for whereas Darwinism as such offends by an unwarranted extrapolation while remaining otherwise faithful to the scientific point of view, theistic evolutionism betrays the theological outlook itself and thereby gives rise to a wholesale corruption of sacred doctrine.

The point needs to be made, in particular, that where there is no vertical descent, there can be no vertical ascent either. A thing which has its first origin in space and time will have its last ending in space and time as well;

such a thing is bound to perish, bound to disappear like a riven cloud. But this is not the case with things that have being, and thus an essence and an act-of-being. Only a cosmology, therefore, which enshrines the dimension of verticality can support a religious outlook and allow a doctrine of human immortality. Within the confines imposed by a horizontal cosmology, the claim of religion becomes a sham, or at best a consoling fiction.

Notes

1. *The Quantum Enigma* (Peru, IL: Sherwood Sugden, 1995), 85-97.
2. *The Christian Philosophy of St. Thomas Aquinas* (University of Notre Dame Press, 1994), 185.
3. Francis Darwin, ed., *The Life and Letters of Charles Darwin* (London: Murray, 1887), vol. 2, 303-12.
4. *Darwin's Black Box* (New York: The Free Press, 1996), 39.
5. *On the Origin of Species* (Harvard University Press, 1964), 189.
6. *No Free Lunch* (New York: Rowman & Littlefield, 2002), 155. It should perhaps be mentioned that the term "no free lunch" has in recent years acquired a technical sense: it refers to a class of mathematical theorems concerning so-called evolutionary algorithms, proved in the late 90's. It turns out that the problem-solving capacity of such algorithms is severely limited. Dembski's theory can be seen as a generalization of these "no free lunch" theorems.
7. See Leon Brillouin, *Science and Information Theory*, 2nd ed. (New York: Academic Press, 1962), 267-69.
8. In point of fact, the proof is not quite as simple as my summary suggests: for example, one must get around the difficulty that the functions f and f_ω may carry CSI, and thus inject CSI into the process. For a full discussion, see William A. Dembski, *No Free Lunch*, op. cit., 149-166.
9. Having discovered the correct radiation formula by empirical means, Planck derived its physical significance by way of a statistical analysis. As he put it in his Nobel Prize address of 1920: "After some weeks of the most intense work of my life, light began to appear to me and unexpected views revealed themselves in the distance."
10. I have reported on this work in "Eddington and the Primacy of the Corporeal," *Sophia*, Vol. 6, No. 2, 2000, 5-38, reprinted here as Chapter 3.
11. *The Reign of Quantity* (London: Luzac, 1953), 13.

12. As the reader may have observed, I have not entered into the mathematical formulation of detachability. Suffice it to say, the condition hinges upon the statistical notion of rejection functions, a technical concept with which we need not concern ourselves in the present summary.

13. *The Quantum Enigma*, op. cit., 106-107.

14. *Opusculum, De unitate intellectus contra Averroistas*, iii; quoted by Joseph Rickaby in *Of God and His Creatures* (Westminster, MD: Carroll Press, 1950), 127n.

15. *Summa Theologiae*, I.117.1 & I.45.6.

16. The question arises whether this conclusion can also be reached on the basis of Dembski's theorem, given that the process in question (the movement of the bow) carries CSI. The problem is this: Conceivably the violinist does no more than transmit CSI, stored presumably in his brain, and thus operates in the manner of a machine. The fact is that Dembski's analysis does not preclude that possibility.

17. I do not wish to suggest that the intellectual act occurs "before" the musical idea is conveyed to the world of sense: my contention is that the intellectual act has ontological as opposed to temporal priority. The intellectual act *can* in fact have no temporal priority, seeing that it is atemporal. It belongs to a realm where there is no "before" or "after," hard as it may be for us to grasp this point. To do so, it seems, we require a metaphysical symbol, a kind of mental icon: think of a circle, for instance, the circumference of which represents the realm of time, of temporal sequence. Inasmuch as the center of the circle is equidistant to all points on the circumference, it is equally "present" to each of these points, regardless of its position within the temporal sequence. What is needed in order to grasp the idea of atemporality is thus an extra dimension: and this is indeed "the dimension of verticality." I might add that it is the loss of that dimension—the inability, in other words, to understand the meaning of metaphysical symbolism—that most accurately characterizes the modern mind.

18. *The Quantum Enigma*, op. cit., 95-97.

19. *Steps Towards Life: A Perspective on Evolution*, (Oxford University Press, 1992), 12.

20. The most compelling exponent of theistic evolutionism, it is safe to say, is none other than Teilhard de Chardin, whom many regard as a veritable prophet. On this subject I refer to my monograph, *Teilhardism and the New Religion* (Rockford, IL: TAN Books, 1988).

21. According to Thomistic ontology, the *ratio seminale* of an organism, strictly speaking, is itself "exterior" to its act-of-being, which constitutes, one may say, the existential basis of that *ratio seminale* itself, and thus indeed the "innermost point" of the organism: the true center of our symbolic circle.

Chapter XI

Interpreting Anthropic Coincidence

What is a man, that the electron is mindful of him?

Carl Becker

By the middle of the twentieth century, physicists had come to realize that the familiar world in which we live and have our physical being constitutes a highly improbable formation, a contingency about as unlikely as randomly picking a premarked grain of sand in the Sahara. By then it was clearly understood that a handful of fundamental numerical constants, such as the ratios of the nuclear, electric and gravitational forces, determine what is physically possible and what is not. One came to recognize that the atomic and molecular structures upon which life depends demand an immensely delicate balance or near-balance of opposing forces, which in turn demands that the constants of Nature assume the values they are found to have with almost a zero margin of tolerance: it is somewhat as if the distance between the centers of the Earth and the Moon, let us say, had to be what it is to within the thickness of a hair. In the face of these recognitions one can hardly refrain from asking *why* the universe should be thus fine tuned. To be sure, the question is more philosophical than scientific, given that the laws and constants of Nature constitute the principles in terms of which scientific explanations are framed. It is not, strictly speaking, the task of science to explain or justify the existence of these laws or the values of these constants; and yet the fact remains that for more than half a century, scientists have debated the issue. An entire literature, claiming at least to be scientific, has evolved in response to this problem.[1]

Although the fact of fine tuning is evidenced by the state of the universe here and now, it happens that the debate has from the start presupposed an evolutionist scenario. Two problems, in particular, have dominated the initial phase of the discussion. The first has to do with the fact that the very large integers and dimensionless ratios arising naturally from atomic physics and the new cosmology have turned out to be close to 10^{40} or its square. For example, the ratio N of the electric to the gravitational force between a proton and an electron is approximately 2.3×10^{39}. Compare this with another exceedingly large number of interest to cosmologists: the present age of the universe in so-called jiffies, where a jiffy is the time it takes light to traverse the diameter of a proton. Now, this quantity, call it M, is calculated to be of the order 6×10^{39}. It was this close proximity between the seemingly unrelated quantities M and N that prompted Paul Dirac (one of the greatest physicists of the twentieth century) to propose, in 1937, what he termed the Large Number Hypothesis: "Any two of the very large dimensionless numbers occurring in Nature are connected by a simple mathematical relation, in which the coefficients are of the order of magnitude of unity." Since M is time-dependent, this implies that N must be time-dependent as well; and this fact led Dirac to propose a second hypothesis: he stipulated that the gravitational constant G is inversely proportional to the age of the universe. On this basis, M and N could remain comparable, and both would presently be large simply because the universe is old. And so the matter stood until 1957, when a Princeton University physicist by the name of Robert Dicke explained the proximity of M and N in a very different way.

In keeping with the big bang scenario, Dicke assumed that the heavier elements upon which life on Earth depends, such as carbon and oxygen, were produced in the interior of a star through nucleosynthesis and released into the surrounding space when the star, having burned up its fuel, turned into a supernova and disintegrated. Now, this process, from start to finish, requires about 10 billion years. Allowing perhaps another few billion years for the formation of planet Earth and the evolution of life, Dicke concluded that it takes somewhere between 10 and 15 billion years for scientific observers to appear on the scene. Taking into account also estimated survival times, he inferred that "The age of the universe 'now' is not random but is conditioned by biological factors." This establishes a close connection between the lifetime of main sequence stars and the constant M representing the present age of the universe. A third large number L, in the form of a stellar lifetime, has now entered the picture; but it happens that L is indeed related to N, given that stellar evolution is determined by the interplay of electric and gravitational forces. Thus, by way of the constant L, Dicke was able to explain, at least to the satisfaction of his fellow scientists, a large number coincidence that had puzzled the astrophysics community for decades. It is to be noted that his explanation brings into play a novel principle, one that came in 1974 to be dubbed "the anthropic principle." Basically, it affirms that the universe must have certain features, because, if it did not, human beings would not be here to observe that universe. In Dicke's case, the point at issue was simply that "carbon is required to make physicists," to put it in his own words. It turns out that this rather obvious fact does enable physicists to hypothesize a close relation between two very large and seemingly very unconnected quantities.

The second problem to which I have alluded has to do with the production of carbon in the interior of a star. It evidently takes three helium 4 nuclei to produce one carbon 12 nucleus. However, since 3-particle collisions are exceedingly rare, one must suppose that the production takes place in two steps: first, two helium 4 nuclei collide to produce a beryllium 8 nucleus, which then collides with another helium 4 nucleus to produce carbon 12. Now, the feasibility of these fusions hinges upon favorable nuclear resonances. In particular, Fred Hoyle ascertained, in 1954, that a carbon 12 nucleus *must have* a resonance level close to 7.7 MeV if nucleosynthesis of carbon 12 is to occur. The following year, while on sabbatical at Caltech, he prevailed upon his nuclear physics colleagues to test this prediction; and as it turns out, a resonance level at 7.656 MeV was in fact detected. Hoyle's unspoken reasoning, one sees, was once again "backwards": from the existence of carbon 12 to its stellar nucleosynthesis, and thus from the present to the past; it belongs implicitly to the "anthropic" genre. To be sure, the story of nucleosynthesis, as it is told today, does not end with the production of carbon, and one finds that a number of other "anthropic coincidences" enter into the production of nuclei "required to make physicists."

However, as I have noted before, the necessity of fine tuning is evidenced, first of all, by the state of the universe here and now. Let us consider a very simple example. The charge of an electron and of a proton as determined experimentally are found to be equal and opposite; their ratio is consequently −1. What would happen, let us ask, if this ratio were ever so slightly changed? Given the atomic constitution of matter, one finds that an explosion would take place. The opposite charges of protons and electrons would no longer cancel out, implying that material objects would become electrically charged (positively or negatively, depending on which of the two elementary charges dominates). Not only, therefore, would objects repel each other, but each would repel itself internally, which is to say, would explode. A change in the ratio –1 by as little as 1 part in 100 billion would suffice to blow our world apart in a split second. Now, this is only the simplest example, one that can he appreciated without any deeper knowledge of physics. Obviously other more refined "anthropic coincidences" are required to permit the existence—as distinguished from a surmised formation via nucleosynthesis—of nuclei, atoms and molecules, as well as of their organic and inorganic aggregates, and to assure the myriad physical and chemical properties of these aggregates upon which our life depends. Remarkably, once nuclei are in place, two fundamental constants suffice in principle to control this multitude of phenomena: namely, the so-called fine structure constant, whose value is approximately .00729720, and the ratio of the proton to the electron mass, which is close to 1836.12. According to quantum theory, a universe thus fine tuned will permit "carbon-based physicists" not only to exist, but to construct the technological apparatus needed to explore the universe.

Clearly, the discovery of anthropic coincidence calls for a shift in the scientific outlook. Scientists had realized for a very long time that living creatures are adapted to the cosmic ambience; but now the realization has dawned that the cosmos itself is adapted to living forms. The adaptation, one finds, is mutual; it now appears that the cosmos, for whatever reason, has been finely tuned for the reception of life, and that the very particles comprising the universe prove to be "mindful of man."

The first enunciation of the so-called Anthropic Principle, one sees in retrospect, had failed to grasp the critical point: "Our location in the universe," Brandon Carter specified back in 1974, "is necessarily privileged to the extent of being compatible with our existence as observers." One senses an effort to accommodate the new insight within the old worldview, a desire, as it were, to circumvent the crux of the matter. Carter's allusion refers clearly to what had come to be known as the Copernican Principle: the notion, namely, that the cosmos, on average, is quite the same everywhere, and that there is nothing at all special, let alone "central," in our cosmic location. In Carter's original version, the Anthropic Principle constitutes thus a complement or correction to the Copernican, in recognition of the fact that the human observer requires a terrestrial environment fine tuned to his exceedingly fragile constitution. Pick a cosmic location at random, and almost with absolute certainty it will prove lethal to man.

Understandably, other versions of the Anthropic Principle were soon to follow, and there exists by now a considerable collection of such formulations, representing various strands of thought. To be sure, all these—with the conceivable exception of the PAP to be discussed below—presuppose the evolutionist scenario, from star-formation to the production, once again, of physicists. The simplest version is the so-called Weak Anthropic Principle (WAP), which Barrow and Tipler formulate as follows: "The observed values of all physical and cosmological quantities are not equally probable, but take on values restricted by the requirement that there exist sites where carbon-based life can evolve, and by the requirement that the Universe be old enough for it to have already done so."[2] The connection with the early work of Dicke and Hoyle is unmistakable. Leaving aside the question what the term "probable" could possibly mean in this context, it is to be noted that Barrow and Tipler have elevated the hypothesis of cosmic and biological evolution to the status of an unquestionable dogma. Immediately after enunciating their WAP, the authors go on to say: "Again we should stress that this statement is in no way speculative or controversial. It expresses only the fact that those properties of the Universe we are able to discern are self-selected by the fact that they must be consistent with our evolution and present existence." There is nothing "speculative and controversial," supposedly, in the stipulation that life on Earth has evolved from nuclei produced out of hydrogen and helium in the interior of stars! Apparently this tenet is no longer a hypothesis, a theory in progress, but has now become a dogma that can serve as the bedrock of our scientific thought.

What proves to be essential in the WAP, however, does not hinge upon any scenario of cosmogenesis; it boils down simply to the logical recognition that the observed universe must be compatible with the fact that it *is* observed. Thus reduced to its essential idea, the WAP is indeed nonspeculative and noncontroversial; and yet, so far from constituting a mere tautology, it entails consequences of great scientific interest. As a self-selection principle, it has been in scientific use for centuries: Copernicus, for example, applied this very principle when he explained the retrograde motion of planets as a self-selection effect resulting from the orbital motion of the Earth. The fact that it is we who are observing the universe—and not some disembodied spirit—does carry weight. It is this principle, precisely, that empowered Dicke to conclude that the observed age of the universe is more than "a brute fact," and it is rigorously true that within the context of the cosmogenetic scenario, that recognition does lead to an understanding of the large number coincidence he set out to explain.

Yet it is also clear that the WAP does not give physicists all that they want. Thus, when Hoyle made his famous prediction, he did so on grounds not ostensibly covered by that principle; the location of a nuclear resonance, after all, is determined by the strength of the nuclear force, and it is difficult to imagine that the fundamental constants of Nature could be affected by a scientific observer, one who, in fact, only came into existence some 15 billion years after the constants in question became operative. Clearly, the logical principle implicit in Hoyle's reasoning was something more than the WAP; it has since been termed the Strong Anthropic Principle (SAP). As formulated by Barrow and Tipler, it affirms that "The Universe must have those properties which allow life to develop within it at some stage of its history." To be sure, the evolutionist hypothesis has once again been incorporated into the formulation; but now, as a component of the SAP, that hypothesis proves to be problematic even to the contemporary scientific mind. It is a fact that the constants of Nature are what they are, and that life exists on Earth; what more, then, does the SAP demand? Is there a difference between "is" and "must be"? It would seem that a determination has somehow been made, a selection in favor of fundamental constants permissive of life. Such a fine tuning, however, at the start of a presumed cosmogenesis, is strongly suggestive of intelligent design, and seems to demand the postulate of a creator. It was to be expected, therefore, that the enunciation of the Strong Anthropic Principle would lead to controversy within the scientific community, and cause divisions within its ranks. A certain minority, comprised of scientists with at least a minimal affinity for theological thought, accepted the SAP at its face value, so to speak; for them the principle assumed the following form: "There exists one possible universe 'designed' with the goal of generating and sustaining 'observers'."[3] Most scientists, on the other hand, having embraced the philosophy of naturalism, have found the notion of a "designed" universe—"designed" even for the purpose of ensuring their own existence—quite unacceptable. The major segment of the scientific community thus found itself confronted by a problem of no mean proportions: how does one explain, in naturalist terms, why the universe *must be* so constituted as to generate and sustain "observers" at some stage of its evolution? However, daunting as the problem may be, one should never underestimate the resourcefulness of physicists. If it turns out—as in this case it does—that a single universe will not do, one overcomes this difficulty by postulating that there are many, and if need be, an infinite number. One can then speak indeed of "probabilities" with reference to fundamental constants, and explain anthropic coincidence by an anthropic concept of self-selection. This option, which apparently has been chosen by a majority of scientists concerned with the anthropic problem, gives rise to another version of the SAP, which Barrow and Tipler formulate as follows: "An ensemble of other different universes is necessary for the existence of our Universe."

Three different ways of generating universes have been proposed; following the suggestive terminology of Karl Giberson,[4] I will refer to these as Recycling, Splitting, and Bubbling. The first kind of generation is said to take place when a big bang-generated universe eventually collapses into a terminal singularity, generally referred to as the "big crunch," and emerges on the other side, figuratively speaking, as a Recycled Universe with duly altered fundamental properties. Needless to say, of course, there is no physics that takes us through that singularity. But be that as it may, continuing in this manner, one obtains formally an indefinite series of universes, from which ours has conceivably been self-selected by the fact that we happen to exist in this particular member of the given ensemble. The second universe-generating scenario, termed Splitting, was first conceived in the context of quantum theory to explain the so-called collapse of the state vector in the act of measurement.[5] To obviate the indeterminism associated with this collapse, a physicist by the name of Hugh Everett proposed the idea that every act of measurement splits the universe into as many parallel worlds as there are possible values of the given observable. To be sure, the observer becomes likewise split—such is the price, it seems, to be paid for the preservation of determinism. In any case, Everett's idea was picked up by physicists concerned with the anthropic problem and applied within that context as a mathematical universe-generating principle. The third method, the idea of Bubbling, was inspired by quantum field theory, which has something to say concerning the production of virtual particles as random fluctuations in a vacuum. As Giberson describes the resultant conception: "Our universe is but one of many embedded in a larger 'meta-universe', out of which new universes are constantly erupting like bubbles from the bottom of a pan of boiling water."[6]

Briefly stated, these are the ways by which physicists of a naturalist persuasion have arrived at a "many-worlds" scenario in their respective attempts to make naturalistic sense out of anthropic coincidence. Basically, what renders anthropic coincidence explicable to the satisfaction of many-worlds theorists is that the problem has now been brought within range of the WAP. The specter of intelligent design has thus been eliminated, replaced by a concept of self-selection. According to the theoreticians, all these many or innumerable universes are equally real, equally significant or insignificant. It appears that in this egalitarian age even universes have been accorded equal rights.

The connection, in particular, with Darwinism should not be overlooked. In fact, the many-worlds approach was suggested independently by Charles Pantin, a Cambridge University biologist, who pointed out in 1965 that "if we could know that our Universe was only one of an indefinite number with varying properties, we could perhaps invoke a solution analogous to the principle of Natural Selection" to explain the occurrence of anthropic coincidence.

I will mention in passing that a third basic version of the SAP, known as the Participatory Anthropic Principle (PAP), has been proposed by the eminent physicist John Wheeler. It too is inspired by quantum theory, and is closely related to the many-worlds approach; its distinguishing feature, however, resides in the fact that it assigns an extraordinary role to the scientific observer: so far from being a mere spectator, it is affirmed that he actually brings the universe into existence! One sees that the level of sophistication of the debate has undergone a quantum jump through the introduction of this idea, and that Wheeler's theory has a distinctively idealist cast, reminiscent of Bishop Berkeley, as Barrow and Tipler point out. It corresponds, in any case, to a third possible interpretation of the SAP, which reads: "Observers are necessary to bring the Universe into being." One wonders, of course, how there could be human observers before there is a universe; but then, as I have said before, we are dealing with a highly sophisticated theory, which cannot be readily disassociated from a technical level of discourse. However, we shall nonetheless have occasion to point out that Wheeler's doctrine is based upon flawed ontological presuppositions.

A fourth version of the SAP—in a way the most interesting of all—known as the Final Anthropic Principle (FAP), states that "Intelligent information-processing must come into existence in the Universe, and, once it comes into existence, it will never die out." The idea derives from a paper by the distinguished physicist Freeman Dyson, which appeared in the *Reviews of Modern Physics* in 1979 under the title "Time without End: Physics and Biology in an Open Universe," and from his Darwin Lecture entitled "Life in the Universe," delivered two years later at Cambridge University. Based upon concepts presented in these papers, a new mathematical discipline has come to birth, presently known as "physical eschatology." According to this theory, living creatures—and especially intelligent creatures—are in effect computers, made up of hardware plus a program. The program, of course, is seen as the scientific equivalent of what formerly was termed the soul. Barrow and Tipler make the point explicitly:

> The essence of the human being is not the body but the program which controls the body; we might even identify the program which controls the body with the religious notion of a *soul*, for both are defined to be non-material entities which are the essence of a human personality.[7]

The hope or intent of eschatology physicists is to show that somehow "programs do not die," which is to say that information-processing can perpetuate itself into future states of the universe, in which conditions will no longer allow the kind of biological hardware which presently exists. We refer the interested reader to Barrow and Tipler's Oxford University treatise, in which the proposed "principle of immortality" is expounded.[8] What confronts us here is a branch of physics, or at least a discipline that claims to be such and exhibits the trappings of a mathematical science, which seems quite overtly to encroach upon the turf of theology: what is one to make of that?

Following this brief glance at many-world and other naturalistic interpretations of the Anthropic Principle, it behooves us to reflect upon these putatively scientific theories. Barrow and Tipler, for their part, characterize their Final Anthropic Principle as "a statement of physics"[9]; but is it really that? The same doubt, to be sure, arises in connection with the various many-worlds theories, not to speak of John Wheeler's Participatory Anthropic Principle. What has happened is that the discovery of anthropic coincidences, when viewed within the context of an evolutionist cosmology, came to be perceived by the scientific community as a problem, and indeed, as a challenge to their naturalistic *Weltanschauung*. Why *should* the electron be "mindful of man"? Why should the fundamental constants of Nature be fine tuned to 1 part in so many billion in order that life may spring into existence in the course of time? Could this be simply an accident? Apart from the fact that one does not quite understand in what sense an entire universe could be "accidental," one has a vague sense that somehow more than accident, more than coincidence, must be at play. But what could that be? Having rejected the theistic interpretation out of hand, the scientist is hard pressed to arrive at an explanation of his own. The Weak

Anthropic Principle seems to be as far as one can go on a rigorous scientific basis; beyond this point the very question is no longer scientific, but becomes perforce metaphysical. Even the doubt—the amazement or sense of wonder aroused by the discovery of anthropic coincidence—proves to be inherently metaphysical. It is in essence the kind of doubt, the kind of wonder, said to be the beginning of philosophy; it constitutes thus an invitation to think in an authentically philosophical mode, and marks a threshold that can lead us, *Deo volente,* beyond the limits of what Shakespeare termed this narrow world. It appears, however, that few scientists have availed themselves of this call, this opportunity, hampered as they generally are by the baggage of scientistic beliefs. As a matter of fact, it is scientistic belief in the form of an evolutionist cosmology, and not "hard" physics, that has brought them to the present impasse; and so the die has been cast. Under these auspices, there seems at this juncture to be no way out but to embark upon the domain of philosophy, not however as a true philosopher, but as a scientist, at least in name.

More than just "hard" physics, I have said, had come into play even before the advent of anthropic theorizing. The transition, therefore, from the original big bang cosmology to Recycled, Splitting, and Bubbling Universes is not quite as radical or discontinuous as one might think, and it is not easy to determine whether the genre of the discipline has remained the same, or whether the border between science, properly so called, and a kind of science fiction has been crossed at this particular turn. One thing is certain: by the time we arrive at a Recycled, Splitting, or Bubbling Universe, the border has indeed been crossed; for it is clear, virtually by definition, that these conjectured universes are situated beyond the pale of scientific observation. There can be no intercommunication between two different universes: no signal originating in one can ever reach the other. Yet, even so, the introduction of extraneous universes cannot really be classed under the rubric of science fiction, inasmuch as the step is taken with serious intent; what stands at issue, after all, is the explanation of anthropic coincidence on a naturalistic basis. One must also remember that the theoreticians who engage in these speculative pursuits are for the most part scientists in good standing, and include highly respected names; we are not dealing here with amateurs or lunatics. Far from it! The level of mathematical expertise and sophistication presupposed by many of these studies is in fact phenomenally high, and the connection with such reputable domains as particle physics and quantum field theory is very close. No, science fiction it is not. A better designation for the emerging genre, perhaps, would be hyperphysics, a physics that exceeds itself, that has transgressed its own proper plane. The point is that physics itself, driven by the genius, one could say, of its practitioners is evolving into something new, something very different from the "hard" science it used to be. An unprecedented proliferation of novel mathematical structures, novel "entities," is in progress, the observational basis for which is becoming ever more tenuous, while at the same time the levels of mathematical abstraction and complexity are reaching unprecedented heights. It is almost as if physics wanted to turn itself into a kind of mathematical metaphysics by virtue of its surpassing abstraction and mathematical prowess; or wants to become, if you will, a theology: a "mathematical theology," no less. It is significant that Barrow and Tipler speak of their FAP as "the physical precondition for moral values"; and again I want to stress that this is something more than simple lunacy. One is in fact reminded of the Teilhardian vision of science turned religion: is this not precisely what is here at stake?

I have noted before that the amazement which seems to have sparked the many-worlds quest is inherently metaphysical. The quest as such is that of science attempting to exceed itself, to transgress its own proper domain, its own inviolable bounds, and turn itself into a veritable "theory of everything," to use a term currently in vogue within the physics community. Now, this drive, this passion, is Promethean whether one speaks of many worlds or only of one. It is illegitimate, moreover, not because it violates some moral commandment, but because it goes against the nature of created things. Everything in creation has its own proper bounds; and science itself is no exception to this rule. I have elsewhere elaborated upon the fact that "hard" science in our day has brought to light a number of what I term limit theorems, which in fact disclose some of these bounds by the methods of science itself;[10] and who can doubt that a limit theorem forbidding such a thing as many-worlds theory also exists in principle. Physics is a science which refers perforce to this universe, the one in which we find ourselves; and in fact, it refers necessarily to a restricted domain or aspect thereof; as Goethe has wisely said: "*In der Beschränkung zeigt sich der Meister.*" The very rigor and exactitude of physics derive from its delimitation. One might go so far as to maintain that the things with which physics is concerned are in fact constituted by the bounds imposed by the *modus operandi* of physics itself: there can be no question, thus, of physics transcending the ontologic plane to which it is geared. To put it another way: physics is able to function precisely because it is *not* a "theory of everything."

Of course, it is always possible to transgress bounds "on paper"; and indeed, one can do so with a semblance of rigor that commands respect. But in truth the resultant structures have perforce lost touch with reality; a science that would cast off its yoke of limitation turns into a pseudoscience in a trice. It may indeed retain the outer form of science, its seeming exactitude, but never its substance, never its content of truth. The structures of such a putative science prove to be empty, no matter how mathematically rigorous, sophisticated, or exacting the founding argument may be; even the celebrated criterion of mathematical elegance is powerless to bestow reality all by itself. One might remark in passing that this is presumably the reason why Albert Einstein produced so very little of lasting interest after 1917, when he published his general theory of relativity, and apparently, from that point onwards, relinquished the method of thought experiments and physical reasoning which had served him so well during his productive years (roughly from 1904 to 1917). Mathematical physics seems now to have entered a similar phase *en masse,* in which the emphasis has shifted from the interplay of theory and experiment to the genre of pure theory, a kind of mathematical universe building, which retains its fictitious character even if the purported universe is identified in theory with our own. Perhaps, in this age of postmodernism, the physicist feels entitled to create his own world, as if truth were nothing more than a mathematical convention.

One more observation. It may be of interest to point out a rather striking analogy between the many-worlds formalism and René Guénon's geometric representation of what he terms the degrees of Existence.[11] It will be recalled that each degree of Existence comes thus to be represented by a horizontal plane in a 3-dimensional (Euclidean) space. Now, if we think of the 3-dimensional space as a meta-universe, and each horizontal plane as a particular world or universe, we have here a representation of the many-worlds scenario, in which there are, quite literally, an indefinite number of parallel universes. But there is more to the analogy: for in Guénon's representation the ensemble of individual beings comprised within a given degree of Existence is represented by a family of parallel straight lines filling the corresponding plane. Now, something similar takes place in contemporary cosmology, where an individual particle, in the integrality of its existence, is represented by a curve in 4-dimensional space-time, its so-called world line. Here too, then, "individual beings" contained within a given world are represented by lines. Of course, we must not forget that the "many worlds" of contemporary cosmological theory, if they were to exist, would be comprised necessarily with a given degree of Existence in Guénon's sense. Nothing, in fact, could be further removed from the mental horizon of a contemporary physicist than the notion of multiple degrees of Existence; for indeed, even within "our world" his *modus operandi* forces him to shrink the ontological spectrum to what may properly be termed a single sub-existential plane.[12] And yet, as has sometimes been pointed out, there is a certain correspondence, an inverse analogy perhaps, between opposites. Could it be, then, that the analogy between the parallel universes of the physicist and the degrees of Existence as envisaged by René Guénon is more than just formal, more than "accidental"? Could the thought of these scientists, anti-metaphysical though it be, have captured nonetheless a distant reflection of a profound metaphysical truth? For my part, I would not rule out that possibility.

Having reflected at some length upon the naturalist response to the puzzle of anthropic coincidence, let us now consider the theistic interpretation, which likewise has assumed a variety of forms. Not every author, certainly, included within this category, is a religious believer. There is the example of George Ellis, a former associate of Stephen Hawking, and now a Professor of Cosmic Physics in Triest: Ellis represents what might be termed the minimalist position within the given camp. He speaks rather guardedly of "a wider framework" of explanation, and concedes that there may be "an underlying structure of meaning beneath the surface appearances of reality, most easily comprehended in terms of deliberate Design," a surmise that might well be described as a first step in the right direction. Again, the well-known physicist Paul Davies has expressed himself in similar terms. Thus, in *Cosmic Blueprint* (1988), he alludes to "an overwhelming impression of design," and four years later, in *The Mind of God,* he writes:

> I belong to the group of scientists who do not subscribe to a conventional religion but nevertheless deny that the universe is a purposeless accident. Through my scientific work I have come to believe more and more strongly that the physical universe is put together with an ingenuity so astonishing that I cannot accept it merely as a brute fact. There must, it seems to me, be a deeper level of explanation.

The sentiment expressed in this remarkably forthright statement of belief is no doubt shared by many serious-minded scientists, and may be about as far as most are willing to go. A comparative few, on the other hand, do "subscribe to a conventional religion." Perhaps the most notable representative of this group is John Polkinghorne, the physicist who resigned a professorship at Cambridge University to become an Anglican priest. Polkinghorne finds the facts of anthropic coincidence to be consonant with the Christian teaching that God created the world. It is not, in his eyes, a matter of proving the existence of God, not an "argument" in the sense of Aquinas, but rather, as he says, a "consonance." To be sure, Polkinghorne accepts the evolutionist scenario—from big bang to origin of species—without reservation, and it is this scenario, precisely, that he perceives as being "consonant" with Christian doctrine. Obviously this entails that he reject the historical interpretation of Genesis (that is to say, of the first three chapters), as do, of course, most contemporary theologians. His understanding of creation, he explains, is ontological as opposed to temporal; the biblical teaching, he maintains, "is not concerned with temporal origin, but with ontological origin. It answers the question: why do things exist at all? God is as much the Creator today as he was 15 billion years ago."[13] The God of Christianity, Polkinghorne insists, is not "the God of the edges, with a vested interest in beginnings," but indeed "the God of all times and all places." Now, this is certainly well said, and basically Augustinian, if you will; the crucial difference, however, lies in the fact that St. Augustine did not simply reject the historical interpretation of the hexaemeron, but transcended it from an authentically metaphysical point of view.

The rift widens, to say the least, when Polkinghorne goes on to explain that "It is in sustaining the fruitful process of the world that God is at work as the Creator." To be sure, "the fruitful process of the world" to which he alludes is none other than evolution as currently understood, which is to say that Polkinghorne is simply repeating the Teilhardian dictum that "God creates through evolution." In keeping with presumed insights of contemporary science, he perceives the "fruitfulness" in question as resulting from "an interplay between two opposing tendencies which we could describe as 'chance' and 'necessity'." And he proceeds to enunciate a theological interpretation of these two principles: "I believe that the Christian God, who is both loving and faithful, has given to his creation the twin gifts of independence and reliability, which find their reflection in the fruitful process of the universe through the interplay between happenstance and regularity, between chance and necessity." A logical point deserves to be made: If indeed God creates by sustaining "the fruitful process of the world" consisting of an interplay between chance and necessity, then it would follow that "the twin gifts of independence and reliability" are not merely reflected in chance and necessity, but must in fact coincide with these factors. In other words, according to Polkinghorne's doctrine of creation, as enunciated above, chance and necessity, so far from constituting secondary principles reflecting a primary duality ("the twin gifts of independence and reliability"), are conceived indeed as the primary duality itself. It would seem that when Polkinghorne speaks of "reflection," he is momentarily falling back to a pre-Teilhardian concept of creation.

A typical example of Polkinghorne's "fruitful process" is furnished by the formation of stars and galaxies, as currently conceived. An originally smooth universe becomes grainy or lumpy by chance, and such small initial deviations become subsequently amplified through the action of gravity, which is where necessity enters the picture. Polkinghorne concludes that "the interplay between those tendencies, chance as origin of novelty, and necessity as the sifter and preserver of the novelty thus produced, is the prime way in which the fruitfulness of the universe is realized." It is to be noted that this is quintessential Darwinism, now applied on a cosmic scale.

It is worth pointing out that in the proposed synthesis of science and theology, what fits the least is the Christian doctrine of the Fall. It will be recalled that Teilhard too had struggled with this problem, and despite his phenomenal powers of imagination, with very little success. "The principal obstacle encountered by orthodox thinkers," he wrote, "when they try to accomodate the revealed historical picture of human origins to the present scientific evidence is the traditional notion of original sin."[14] It seems that Polkinghorne is no exception to this rule. In *Reason and Reality,* for instance, he deals with this issue in the final chapter, comprising exactly five pages. He does not attempt to interpret the Fall on an evolutionist basis, but is concerned rather to explain why the doctrine proves to be problematic from a scientific point of view. As might be expected, he insists that "Genesis 3 is to be understood as a myth about human alienation from God and not as an aetiological explanation of the all too evident plight of humanity."[15] One might of course raise the question first posed by Teilhard de Chardin: "Is this still Christianity?"; but that is another matter. What concerns us at the moment is the fact that even when it comes to scientists perceived to be Christian believers, it appears that the presumed evidence of science carries greater weight by far than the traditional teachings of the Faith: wherever a seemingly irreconcilable conflict exists, it is invariably the Christian dogma that loses out.

It is to be noted, in particular, that the very notion of sin, and indeed, of responsibility, has no more place in an evolutionist universe, where freedom must be read as "chance;" for clearly, in chance there is no responsibility, no moral good or evil. The mystery of human freedom—of what is traditionally referred to as the freedom of the will—lies deeper by far than the notion of chance. It has to do, not with matter, but with the opposite ontologic pole, which has been left out of consideration by the Darwinists. True freedom, as I have argued elsewhere,[16] enters the world by way of vertical causation: a mode of causality which acts above time, and which for this reason can find no place in an evolutionist cosmology.

Freedom is neither chance nor necessity, nor a combination of the two, but something that exceeds the plane of natural causation, and which consequently cannot be understood in scientific terms. But clearly, a theology without even the concept of human freedom and moral responsibility is not in truth a theology at all; such a doctrine can be no more than a poor imitation, a clumsy Ersatz.

Yet it happens that Polkinghorne's outlook is defective even from a scientific point of view. In light of intelligent design theory, one now knows that the twin principles of chance and necessity do not suffice to account for "the fruitful process of the world." It turns out, speaking in contemporary scientific terms, that "fruitfulness" translates into complex specified information or CSI, and this is precisely what chance and necessity, singly or in combination, can *not* produce.[17] The most striking example, to be sure, is provided by the "tons" of CSI in the DNA of every living creature, from a bacterium or ameba to the human organism, not one of which could in truth be produced by a stochastic process. If, therefore, one assumes that existing species have evolved by way of common descent, one is obliged to posit acts of vertical causation to account for the production of the requisite CSI. Presumably these interventions from above will be conceived as more or less localized at the branch points of the stipulated genealogical tree, in keeping with the idea of "saltations" espoused by neo-Darwinists such as Richard Goldschmidt and Stephen Jay Gould. In one way or another, vertical causation *must* come into play, even from a strictly scientific point of view. But it appears that Polkinghorne is not privy to this fact, which only became known in 1998.[18] Little did he realize, when he lectured in 1996, that the quintessential Darwinism, which he not only accepted without question but elevated to a theological status, would be rigorously disproved within two years.

Given that an act of vertical causation on a cosmic scale is tantamount to an act of God, one sees that God does once again enter the picture. It appears that the ancient "argument from design" does after all carry validity. With the mathematical discovery of what Bill Dembski refers to as "design inference," natural theology in the sense of William Paley has been revived. By way of a curious dialectic, a materialist science, which seemingly had banished God centuries ago, has now arrived at an impasse which only that "useless hypothesis" can break. By the same token, however, one sees that the God of evolution—or better said, the God of evolutionists—is perforce a "God of the gaps," whose function it is to produce the requisite CSI at critical junctures where stochastic process fails. Yet the fact remains that a God of the gaps proves ever to be an unfelicitous notion: a matter invariably of poor science begetting bad theology. It is somewhat as if a mathematician, finding that the two sides of an equation do not balance, were to add a "God term" to make up the difference: far better to go back to the drawing board and check one's calculations. In the present instance, the correct scientific response to the impasse, it would appear, is to abandon the Darwinist hypothesis of common descent; from a strictly scientific point of view, common descent proves not to be possible. Why should a scientist invoke God to rescue a failed theory: what kind of science is that?

To be sure, a theistic reading of anthropic coincidence is by no means unjustified; the idea that the universe has been designed to serve as a habitat for man is after all traditional. The problem, however, with contemporary theistic interpretations of anthropic coincidence resides in the fact that they presuppose the evolutionist scenario. It thus comes about that the God alluded to is indeed the God of evolution, the God who presides over the stipulated evolutionary process. One role of this God, evidently, is to fine tune the cosmos; it is he who supposedly adjusts the fundamental constants so that carbon 12 will have a nuclear resonance at 7.656 MeV. This kind of theology (if one may call it such) seems to legitimize all manner of scientific theories and hypotheses by conveying the impression that these are literally God-given facts. Who, for example, can argue with a quantum field theory, let us say, put in place by none other than God? The new collaboration between science and theology proves thus to be a boon to both sides: for the theologian it means that at long last he will again be taken seriously, whereas for the scientist it entails that the prestige of God himself will now rub off on his theories.

Whether we realize it or not, theology itself is undergoing a profound transformation; the new affiliation with science is having its effect. Brand new theologies are coming into vogue. I remember attending a symposium at which a Catholic Archbishop delivered a lecture having to do with the work of Stephen Hawking, and confided, during the ensuing discussion, that he personally prefers process theology to Trinitarian doctrine. What surprised me the most was that not one person among the assembled scientists, theologians, and priests, seemed to be in the least astonished or taken aback by this amazing revelation. I came away with a keen sense that Teilhard de Chardin had triumphed at last, that his life-long preoccupation—"the effort to establish in myself and to spread around a new religion (you may call it a better Christianity)"[19] —was bearing bountiful fruit. Whatever one may think of the Teilhardian doctrine, it must be admitted that his vision of a new religion "burgeoning in the heart of modern man from a seed sown by the idea of evolution"[20] was nothing less than prophetic.

From a traditional perspective the phenomenon of anthropic coincidence comes to be perceived in a very different light. One sees immediately that for all their diversity, contemporary approaches to the anthropic problem share the common assumption that the corporeal world—from the planet Earth to living organisms—has been produced through an aggregation of particles. But clearly, what results from an aggregation of particles *is* an aggregation of particles; and thus one assumes in fact that corporeal being "is made of atoms," as every schoolboy nowadays believes. The whole has thus been reduced to its minutest parts; and it is worth pointing out that in view of the Aristotelian definition of quantity as that which admits mutually external components, that reduction places corporeal being squarely in the domain of quantity. This, then, is the ontologic premise which has been assumed throughout the length and breadth of the anthropic debate, by theists and naturalists alike, by the proponents of design no less than by the advocates of chance.

Now, the first thing that needs to be done, clearly, if one wishes to examine the issue at hand from an authentically metaphysical point of view, is to discard the aforesaid premise. It then becomes evident that the debate on record, regardless of the individual positions assumed, has been wrong-headed, a matter of putting the cart before the horse. From a traditional point of vantage one sees that the whole comes first, and then the parts; and if it turns out that the assemblage of parts is somehow "fine tuned," this can only be due to the fact that the parts derive from the whole. The very idea that God adjusts the laws and constants of Nature—that he "monkeys with physics" as Fred Hoyle once put it—to produce the organic world turns out to be ill-conceived and backwards. What God does, theologically speaking, is to *create*; and what He creates—*ex nihilo* and *omnia simul*!—are not quantum particles, but *beings*. Particles come later, ontologically speaking, which is to say that they constitute a secondary reality; and as I have argued at length in my article on "Eddington and the Primacy of the Corporeal,"[21] it is in a sense man, and not God, who "makes" these particles. John Wheeler was right: we do find ourselves in a "participatory universe"; only it needs to be understood that the universe to which Wheeler refers is the physical as distinguished from the corporeal, the existence of which he seems not to have recognized.

One sees that the picture has changed drastically, and has become in a way inverted; the following schematized example may help to make this clear. Consider a 2-dimensional "universe" in the form of a trapezoid ABCD with base AD, and suppose that astrophysicists residing in this universe have arrived at a big bang cosmology. Having ascertained the coordinates of the sides AB and CD, and calculated the coordinates of the point O at which the corresponding lines intersect, they have hypothesized that ABCD must have originated as a rectilinear emanation from that point of origin. Let us consider now the angle ø subtended by OB and OC. From the standpoint of the local scientists, an anthropic coincidence has now come to light: "If the initial angle of dispersion had differed from ø by so much as a billionth of a degreee, *we* would not exist." To the resident naturalists this suggests that there must be an infinite number of parallel universes, presumably in the form of trapezoids, corresponding to all possible values of the initial angular dispersion, out of which the present universe has been self-selected in accordance with the WAP; to certain other inhabitants of ABCD, moreover, it means that ø was determined from the start by an act of intelligent design. In reality, however, it turns out that the triangle OBC constitutes what I have elsewhere termed "the extrapolated universe," and that the angle ø is determined, neither by chance nor by providence, but by the given trapezoid ABCD. What stands at issue can be stated quite simply as follows: The actual universe determines the extrapolated universe, constants and all. One sees that the evolutionist premise has inverted the causal nexus between these two portions of the intentional universe.

It needs however to be admitted that the evolutionist contention is not easily refuted on a scientific plane, and that a number of puzzling questions remain for which no ready answer appears to be at hand. Consider, for example, Fred Hoyle's prediction of the celebrated nuclear resonance in carbon 12. Given that carbon 12 nuclei exist and can only be produced by way of nucleosynthesis in the interior of stars, Hoyle had argued that whatever nuclear resonances are required to permit such a synthesis must in fact be present; and as it turned out, his prediction was verified. One knows, of course, that an experimental confirmation does not suffice to guarantee the truth of a theory: that even if all nuclear resonances required to account for the stellar provenance of chemical elements are found to exist, the claim of stellar origin would still be subject to doubt. Yet it is likewise clear that the evidence in question does favor the evolutionist hypothesis. To be sure, nuclear resonances are determined in principle by the fundamental laws and constants of physics; however, from a non-evolutionist point of view, there is no apparent reason to suppose that the relevant constants should be fine tuned so as to permit a whole series of nuclear reactions by which all elements required by living organisms can be built up, step by step, out of hydrogen and helium. I say "apparent" reason, because it is quite conceivable that as our knowledge deepens, a physical explanation of this fact may come to light; the history of science, after all, provides countless examples of what at first appeared to be a fortuitous state of affairs, but which later could be explained in scientific terms. At the present time, however, we seem to have no such explanation for the nuclear resonances in question; we have only the anthropic argument employed by Hoyle, which presupposes the evolutionist hypothesis. As the matter stands, there is evidence both for and against the big bang scenario,[22] and it is difficult, from a scientific point of view, to gauge the present status of the theory. One may surmise that in the case of each individual scientist it is finally a philosophic leaning that tips the balance. What is more, I contend that the issue cannot in fact be resolved on strictly scientific ground, which is to say that the question is incurably philosophical. What is called for is the ontological distinction between the corporeal and the physical, the primary and the secondary reality at stake; and once this crucial discernment has been made, it becomes apparent that the cosmogenetic scenario as envisaged in big bang theory is ontologically impossible: a secondary reality cannot give rise to the primary, nor can it antedate the primary, except of course in the sense of a pre-existent potency. One sees, moreover, that by its very *modus operandi*, physics is debarred from ever addressing the cosmogenetic issue, and that any cosmogenetic scenario conceived in physical terms is *ipso facto* illusory.

It remains now to reflect upon the significance of anthropic coincidence from the non-evolutionist point of view at which we have arrived. What should we make of the fact that the fundamental constants of the physical universe are fine tuned so as to permit, not exactly corporeal being, but related subcorporeal structures, what I have elsewhere designated as the physical object SX associated with a corporeal object X?[23] Now, the first major point to be made is that the physical universe is in this sense permissive of the corporeal precisely because it is in truth derived from the corporeal domain: by virtue of its secondary or derived ontological status it is logically impossible that the physical universe should *not* be thus permissive of the corporeal. If you break a clay pot, you will find that the resultant shards fit together perfectly so as to constitute the pot in question; and obviously this "fine tuning"—which seems quite miraculous so long as one does not know the true provenance of the shards—is the result neither of chance nor of design. In short, the physical universe is fine tuned because the corporeal world demands as much. It is to be noted, moreover, that the laws and fundamental constants of physics are determined by the process of measurement, as Eddington had foretold, and as Roy Frieden appears now to have confirmed by way of his information-theoretic analysis.[24] The fine tuning of the physical universe can thus be seen as a signature attesting its corporeal provenance by way of mensuration.

But the question remains: How does man, the *anthropos*, enter the picture? If the Anthropic Principle has to do simply with the fact of corporeal primacy, one should drop that designation and speak instead of a Corporeal Principle. However, we need now to take a closer look at the process of measurement. In the first place it is to be noted that a scientific instrument is more by far than mere corporeal matter, which is to say that it constitutes an engineered artifact. Something absolutely essential, obviously, has been added by man. To put it in contemporary terms: it is the scientific observer himself who contributes the complex specified information or CSI to which Nature responds in the act of mensuration; and as Roy Frieden has demonstrated, the laws and constants of physics arise from that interaction. But not without the intervention of another factor: the *theoretical* factor, that is: "Whatever we have to apprehend," notes Eddington, "must be apprehended in a way for which our intellectual equipment has made provision."[25] Here too, on this higher and distinctly "anthropic" level, a process of selection and formation takes place, which proves to be essential to the economy of measurement. Thus, so long as we employ Eddington's metaphor of the net—an excellent metaphor, to be sure—we need to bear in mind that the "net" in question has both a corporeal and an intellectual component. One sees that the physicist functions in the fullness of his trichotomous nature, much like the artist who works externally with chisel or brush, and internally "through a word conceived in his intellect," as St. Thomas Aquinas says.[26] Here too one finds that a "word"—an element of logos—comes necessarily into play. It needs however to be understood that the scientist does not stand by himself, but "participates" in the primary Logos, "the Word that was in the beginning": truly, "Without Him was not made anything that was made." Even secondary realities cannot be "made without Him": it is by virtue of the aforesaid "participation" that the scientist derives his creative prowess, his ability to bring secondary realities into existence. As we have come to see, he too is a "maker," an artist of a kind, and like every true artist, he acts "in imitation of *natura naturans*," the primal Artist who is none other than God.[27]

Now, this insight sheds light on many things. It explains, in particular, what Albert Einstein has termed "the most incomprehensible thing about the universe": the fact, namely, that the universe proves to be comprehensible. The physical universe is comprehensible to the scientist for exactly the same reason that a work of art is comprehensible to the artist; it turns out to be mathematical because it was in fact "made" by mathematical physicists: "The mathematics is not there till we put it there,"[28] declared Eddington to the amazement of his colleagues. As the British astrophysicist said long ago: "We have discovered a footprint in the sand; and lo! it is our own." Here at last we have it: in this decisive recognition lies the ultimate significance of the Anthropic Principle, its burden of truth.

Notes

1. See especially the now classic treatise by John D. Barrow and Frank J. Tipler, *The Anthropic Cosmological Principle* (Oxford University Press, 1986).
2. Op. cit., 16.
3. Op. cit., 22.
4. "The Anthropic Principle," *Journal of Interdisciplinary Studies* 9 (1997), 63-90.
5. I have dealt with this question in Chapter 3 of *The Quantum Enigma* (Peru, IL: Sugden, 1995) from a traditional metaphysical point of view. See also "From Schrödinger's Cat to Thomistic Ontology," *The Thomist*, 63 (1999), 49-63; reprinted here as Chapter 2.
6. Op. cit., 74.
7. Op. cit., 659.
8. See also Frank J. Tipler's later book, *The Physics of Immortality* (New York: Doubleday, 1994).
9. Op. cit., 23.
10. "Science and the Restoration of Culture," *Modern Age*, 43, 1 (2001), 85-93; reprinted here as Chapter 12.
11. *Symbolism of the Cross* (London: Luzac, 1958), Chapter 11.
12. Regarding the "sub-existential" character of physical reality, see especially my article "*Sophia Perennis* and Modern Science," first published in *The Philosophy of Seyyed Hossein Nasr* (La Salle; IL: Open Court, 2001), and reprinted here as Chapter 1.
13. "So Finely Tuned a Universe," *Commonweal*, August 1996, 11-18. Unmarked quotations by Polkinghorne are taken from this article.
14. *Christianity and Evolution* (New York: Harcourt Brace Jovanovich, 1971), 36.
15. *Reason and Reality* (Hauppauge, NY: Trinity Press International, 1991), 73.
16. "Intelligent Design and Vertical Causality," *Sacred Web*, 10 (2002), 49-71; reprinted here as Chapter 10.
17. See my article "Intelligent Design and Vertical Causality," op. cit..
18. William A. Dembski, *The Design Inference* (Cambridge University Press, 1998).
19. *Lettres à Léontine Zanta* (Paris: Desclée de Brouwer, 1965), 127.
20. *Activation of Energy* (New York: Harcourt Brace Jovanovich, 1970), 383.
21. *Sophia*, 6, 2, (2002) 5-38; reprinted here as Chapter 3.

22. Regarding the evidence against big bang cosmology I refer to my article "The Pitfall of Astrophysical Cosmology," *Sophia,* 7, 2 (2001), 5-27; reprinted here as Chapter 7.
23. See Chapter 2 of *The Quantum Enigma,* op.cit..
24. *Physics from Fisher Information* (Cambridge University Press, 1998); see also Chapter 3.
25. *The Philosophy of Physical Science* (Cambridge University Press, 1939), 115.
26. *Summa Theologiae*, I.117.1.
27. I am alluding specifically to the Scholastic doctrine of art; see for instance Ananda Coomaraswamy, *Christian and Oriental Philosophy of Art* (New York: Dover, 1956).
28. Op. cit., 137.

Chapter XII

Science and the Restoration of Culture

My first point is scarcely controversial: From the eighteenth century onwards, I maintain, science has been the major determinant of culture in the West. The influence may be direct or mediated, and the response affirmative or oppositional, but the fact remains that in every cultural domain science has played a pivotal role as the prime agent of change. Take philosophy or theology, social or political norms, art, morals or religious practice: the story is the same. Like it or not, science is the decisive factor—the great new revelation—to which society at large has for long been reacting in multiple ways. Even as technology, the offspring and partner of science, has radically transformed the outer life of Western civilization, science itself is having its impact upon our inner life: upon our basic beliefs, values and aspirations. Not everyone, of course, has become an outright materialist; but all, I submit, have been profoundly affected nonetheless.

From its inception the new science has prospered visibly, and commended itself within ever-widening circles as the great liberator from ignorance and superstition. The age of Enlightenment was upon us, and in a very real sense, still is. Was not Bertrand Russell speaking for the modern world as such when he declared: "What science cannot tell us, mankind cannot know"? An exclusive faith in science appears indeed to be the hallmark of modernity.

That faith itself, however, has begun to falter: we have entered the era of postmodernism. It is not simply a matter of one worldview triumphing over another, as has happened in the past. The shift to postmodernism is far more radical than that; for it denies the validity, not just of an antecedent worldview, but of worldviews in general. Truth has been reduced in effect to a social convention, the local construct of a society. Partly in reaction, no doubt, to the tyranny of the scientistic *Weltanschauung*, one has set about to relativize *all* worldviews. What confronts us here, moreover, is not simply a philosophic trend, but a cultural phenomenon: a cultural revolution, one can say. Think of the wholesale rejection of traditional norms, the pervasive distrust of authority, the radical disorientation which seems especially to afflict the youth of our day. There are of course notable exceptions and indeed counter-trends; but these do not offset the nihilistic tendencies in questions. One has reason to believe, moreover, that there is a real connection between postmodernist philosophy and corresponding cultural trends, even if it may not be possible to construe that connection as a simple case of cause and effect. One can therefore speak of postmodernism in a broad sense, which includes its cultural manifestations.

What I wish now to point out is that postmodernism is not simply an oppositional reaction to the antecedent modernism, but is in fact implicit in modernity, that is to say, in the scientistic worldview itself. The universe as depicted by modern science is clearly unacceptable as a human habitat. The scientistic *Weltanschauung* is bearable, thus, precisely because no one believes it—I mean, fully, with all his being. We believe in the scientistic universe with a part of our mind, persuaded that the contention has been validated by rigorous scientific means; and yet we still suppose, in our daily lives, that the grass is green and the sky blue (which scientism denies), not to speak of the fact that we take a man or a woman to be more after all than a "chemical machine." We have learned to compartmentalize our beliefs: to pass in a trice from one persuasion to another, incompatible with the first, and think nothing of it. This way of managing beliefs needs of course to be learned; it is what modern education has done for us. The art is acquired in schools and universities. The practice, to be sure, is astonishing, if only one stops to think of it; but we generally don't. We have learned the art so well that we are hardly conscious of doing anything at all. As is the case in schizophrenia, we are unaware of our own inconsistency, until of course we engage in authentic philosophical reflection; but even then we rarely perceive the magnitude of the dilemma. It takes a Kierkegaard or a Nietzsche, apparently, to become profoundly disturbed. For most of us the anguish is potential rather than actual, it seems.

It appears from these sparse indications that postmodernism is latent in the scientific mentality. To oscillate between two contradictory worldviews is to commit to neither: to commit to nothing at all. As a chronic condition the practice is tantamount to a denial of truth.

Given that the scientistic outlook is humanly untenable, it behooves us to ask whether the *Weltanschauung* in question is essential to science as such. As one knows, modern science began as an amalgam of Cartesian metaphysics and Baconian empiricism, the incongruity of which was spotted soon enough by leading philosophers. The union, it turns out, is not a true synthesis, and what matters, in fact, is not the Cartesian ontology, nor its epistemology, but precisely the Baconian method. It is Bacon's *novum organum*, his "new machine for the mind," that enables the enterprise of modern science, a science in which "human knowledge and human power meet in one" as Bacon had foretold. To be sure, the Cartesian conception of *res extensa* (of "bare matter") has played a vital role in the motivation and guidance of scientific inquiry. As a Kuhnian paradigm, however, the notion of a clockwork universe is expendable; it is not an essential of science, but only a transitional aid. What ultimately counts, I say, is the methodology, the Baconian character of the enterprise.

The primary reductionism of science is thus methodological; it applies, not to reality as such, but to the means by which we propose to grasp and harness reality. Directed as it is to the objective of control, the Baconian enterprise is inherently designed to count, measure, and quantify; nothing in fact fulfills the Baconian guidelines more perfectly than a mathematical physics. This methodological reductionism, however, does not presuppose, nor entail, an ontology; it is metaphysically neutral, one can say. But whereas science does not *de jure* authorize a reductionism of the ontologic kind, it does so *de facto*; as a rule the tendency to deny what science cannot grasp proves irresistible. It is a fact that science begets scientism, and bifurcation is doubtless the primary scientistic dogma. As Gilbert Durand has wisely observed: "Dualism is the great 'schizomorphic' structure of Western intelligence."

It goes without saying that this "structure" of Western intelligence implicates a worldview and indeed a culture. I concur moreover with Theodore Roszak that "there are never two cultures; only one—though that one culture may be schizoid." And such indeed is our predicament. It should of course be added that not everyone living within the ambience of that culture is *of* that culture—but this is another matter. For better or for worse, there *is* a Western culture, even as there is a Western worldview; and both derive their support from Western science. Certainly that support is illegitimate; yet in terms of effectiveness, this fact carries no weight at all.

Having spoken of bifurcation as the primary scientistic dogma, I should point out that there exists a plethora of secondary scientistic dogmas which hinge upon the primary. Take Darwinism, for example: in response to those who think of that doctrine as a well-substantiated scientific theory I will refer to the growing scientific literature which proves that it is not.[1] Darwinism, it turns out, has never met the Baconian criteria of scientific legitimacy. However, given bifurcation—plus the associated idea that the universe consists of atoms or fundamental particles moving to no purpose, whether by chance or in accordance with deterministic laws—given this reductionist scenario, I say, there *is* basically no other way of conceiving biogenesis and speciation. And this, one may surmise, is the principal reason why scientists continue to cling to some form of Darwinism despite its astronomical improbability. Darwinism, thus, is to be ranked as a scientistic dogma. But whereas this particular dogma constitutes evidently one of the prime examples of scientistic belief, our textbooks are filled with tenets no less spurious, which likewise claim the status of a scientific truth. It suffices that these tenets fit into the prevailing worldview and lend a kind of mutual support; the fact that they do not pass scientific muster remains generally unrecognized, and perhaps would not cause too much consternation if it did become known. It is only when a crisis arises in a particular field that scientists—some scientists, at least—become motivated to engage in foundational inquiries; and even then, it appears, a clear-cut discernment between scientific fact and scientistic fiction is rarely achieved.[2]

One sees that assumptions of a philosophic nature, as well as ideological commitments, do affect the scientific enterprise, which in fact is not quite as "scientific" as one tends to suppose. Scientists are human, after all, not robots or computers; and the postmodernist philosophy of science does after all have a point. And yet, surprisingly perhaps, there *is* such a thing as "hard" science: a rigorous discipline capable of real discovery. Such science carries its own exactitude which no man can bend, and discloses objects or theorems which—like Mount Everest—are simply *there*. Hard science, it turns out, is wiser in certain ways than the scientists who engage in its pursuit, and wiser in a sense than the society sponsoring the enterprise. With a single decree it can abolish a long-standing expectation or disqualify some hallowed canon of scientistic belief; it has in a very real sense a life of its own, independent of social conventions, philosophic bias, or ideological orientation. Apart from technical competence and occasional genius, it demands just one thing from the scientific community: integrity, namely, a certain respect for truth. And happily it can be said, to the honor of that community, that its members have by and large proved worthy of this trust.

It needs now to be pointed out that something momentous and utterly unexpected has taken place within the scientific domain in the course of the twentieth century: science has begun at last to discern its own inherent limitations, its own categorical bounds. Not that it has disavowed its exactitude: not at all! What science *has* disavowed is the scientistic notion that these exactitudes apply in principle to every domain, that *de jure* science encompasses all truth. It has moreover arrived at this recognition of its own incapacity, not by way of some supra-scientific intuition, but by strictly scientific means. What stands at issue are indeed theorems, discoveries as inexorable as the certitudes of mathematics or the fundamental laws of Nature. I propose now to cite a few major examples of such twentieth-century "limit theorems," spanning the gamut from mathematics and physics to biology and cognitive psychology; the cultural implications of these discoveries will occupy us later. For the moment it suffices to note that these remarkable findings are supportive of my "absolutist" claims in behalf of what I have termed hard science.

I will cite, as my first example, the Incompleteness Theorem established in 1931 by Kurt Gödel, a 25-year-old Austrian mathematician, which arguably constitutes the most important discovery of a logical kind in the twentieth century. Gödel's theorem disqualifies, at one stroke, the long-held expectations of leading authorities, the likes of David Hilbert, Gottlob Frege and Bertrand Russell, who thought that a formal system inclusive of all mathematical truth could be found. What the young Austrian proved—once and for all!—is that a consistent formal system rich enough to accommodate ordinary arithmetic is necessarily incomplete; there simply *is* no formal structure encompassing all mathematical truth. It may be noted that Gödel's theorem has a certain postmodernist ring: by restricting the scope of a single theory, a single formal system, it seemingly opens the door to a pluralist outlook tolerant of alternative positions. But even so, it does not compromise the absolute claims of truth: Gödel's result, after all, is a theorem of mathematical logic, validated by a rigorous argument, an incontrovertible proof. It does not in any way relativize mathematical truth; and one might add that Gödel was personally a Platonist, worlds removed from postmodernist skepticism.

My second example has to do with quantum theory, which can be viewed as entailing a limit theorem of a very different kind. What first comes to mind is the Heisenberg Uncertainty Principle, which limits the accuracy with which the values of conjugate dynamic variables of a quantum system (such as position and momentum) can be ascertained. What stands at issue, as one believes today, is not simply an incapacity on the part of the experimentalist, but the fact that dynamic variables of a quantum system do not, in general, *have* a definite value. An electron, for example, may not have a definite position or a specific momentum, and in any case, can never have both at once. It follows that Heisenberg Uncertainty restricts the applicability of pre-quantum physics to a macroscopic domain within which quantum effects can be neglected. And this is one way in which quantum theory can be seen as entailing a limit theorem.

On closer examination, one finds however that more is at stake; as I have shown at length in the first three chapters, quantum mechanics entails an ontological distinction between the physical and the corporeal domains. To be precise, it is the phenomenon of state vector collapse that demands what may be termed the rediscovery of the corporeal world, a domain which in principle falls beyond the reach of physical science as such. What confronts us here is a limit theorem so radical and so profound, that physicists have as yet been unable to come to grips with its implications. Yet this is not all: as I have shown in Chapter 4, quantum mechanics entails in addition a limit theorem which affirms the boundedness, not only of the physical domain, but of the space-time continuum itself, which encompasses the corporeal world. In light of traditional ontology, this is tantamount to a rediscovery of the intermediary domain. I cannot but agree with Henry Stapp that this constitutes indeed "the most profound discovery of science": for it takes us, ontologically speaking, to the deepest stratum of cosmic reality which science can detect.

Another example of a limit theorem is afforded by William Dembski's recently promulgated theory of design, with which we have dealt at length in Chapter 10. Dembski proves that neither chance, nor necessity, nor the combination of the two, covers the entire ground of causality. He demonstrates in particular that what we normally recognize as instances of design cannot be ascribed to the aforesaid modes of causation. These findings, clearly, constitute a causal limit theorem, which imposes stringent bounds upon the domain of phenomena that can be accounted for on the basis of natural causation, and can thus be explained in scientific terms. It happens, moreover, that the entire biosphere, from viruses on up, lies on the yonder side of this divide, which is to say that the origin of biological structures cannot be ascribed to natural causes. If quantum theory has brought to light the existence of higher ontological planes, Dembski's theorem demonstrates that neither the physical universe nor the corporeal (nor the two together) constitute what scientists term a closed system.

My last example concerns the theory of visual perception proposed by the late James J. Gibson, a Cornell University psychology professor who devoted fifty years of his life to the study of how we perceive.[3] By way of painstaking empirical investigations he became convinced that prevailing theories of visual perception are in fact untenable. The very notion that "the eye sends, the nerve transmits, and a mind or spirit receives" needs to be radically modified. In the final count, perception is to be conceived as an act, not of the body, nor of a mind, nor indeed of the two operating in tandem, but of the mind-body compound, conceived holistically as a single entity. What Gibson terms the perceptual system is not a sum of parts, nor can the perceptual act be dichotomized into stimulus and response. And as to the famous "perceptual image"whether conceived as existing physiologically in the brain or psychologically in the mind—he concludes that the concept is spurious.[4] What is perceived, Gibson finds, is not an image, but quite simply the external environment; in a word, the so-called *ecological* theory of perception is non-bifurcationist. "This distinction between primary and secondary qualities is quite unnecessary," writes Gibson, and is in fact "wholly rejected" in his approach.[5] One is amazed to see how this sober scientist was able, by way of hard-headed inquiry based squarely upon empirical findings, to deconstruct the Cartesian edifice. He shows that the customary neurological and computer-theoretic approach to perception is flawed, and can at best yield results of a secondary nature—a recognition which can be seen as a decisive limit theorem pertaining to cognitive psychology.

Hard science, it turns out, is ultimately destructive of scientistic myth. The scientific enterprise is inherently self-corrective: hard science, as I have said, is in a sense wiser than the individual scientist. Consider the phenomenon of state vector collapse, for instance: this finding came, not only as a complete surprise, but as a shock to the scientific community. Erwin Schrödinger, one of the founders of quantum theory, was so disturbed by this phenomenon—"this damned jumping," as he called it—that he rued his own discovery. One sees, in particular, that the limit theorems of quantum mechanics—or better said, the findings which underlie these theorems—have thrust themselves upon the scientific community by force of an inexorable logic, which not even the likes of Albert Einstein could thwart. I can conceive of no more telling proof that such a thing as "hard science" does exist.

We need also, however, to note the following: If hard science is indeed wiser than the scientists, it is wiser too than the postmodernist philosophers who impugn the enterprise. I am not denying for a moment that the new philosophy of science has contributed major insights: scientific facts, for example, may indeed be "theory laden," as I have noted before, and this is of course important. What troubles me about the postmodernist outlook, on the other hand, is its pervasive relativism, which attacks truth the way an acid eats up metal. To be sure, this is hardly the place to attempt a critique of postmodernism; let me however note that Frithjof Schuon may in fact have hit the nail on the head in observing that "its initial absurdity lies in the implicit claim to be unique in escaping, as if by enchantment, from a relativity that is declared alone to be possible." Meanwhile science is continuing to evolve, continuing to unfold its possibilities, and has now attained levels of discovery that shake the foundations of scientistic belief. Hard science, I contend, has disavowed the very premises that gave rise, first to modernist enlightenment, and two centuries later, by way of reaction, to postmodernist skepticism.

What then? If the postmodern, as well as the modern outlook, have now been discredited or found wanting, what viable option remains? As Alexander Solzhenitsyn once declared before an incredulous audience: the only way left today is "upwards." But this way, surely, cannot be improvised, drawn out of thin air: it needs rather to be rediscovered, appropriated, *received.* The very concept of "verticality" is foreign to modernity and postmodernity alike. It reaches us from a distant past, a past we have been taught to despise as primitive and superstitious. We have forgotten, most of us, that tradition can be more than a custom, a convention, a mere vestige to be discarded at will. That a tradition can be living and life-giving, that it can even transmit an element of revelation—this we find hard to believe. We do not think highly of human culture in this Darwinist age. My point is simple: It is time to pay serious and respectful attention once more to the great traditions of mankind; instead of turning postmodernist, let us rejoin the greater human community.

What is called for is a rediscovery of traditional cosmology. Culture and cosmology, as I have said before, go hand in hand; and what is lacking in the modern West—in our culture and in our cosmology alike—is precisely the dimension of verticality. As our universe flattens, so does our conception of man, and so does our culture in all its aspects.[6] There are compensations, of course, on various horizontal planes; but these do not suffice: man was not born for that. He needs the vertical dimension to be fully human. There are those, to be sure, who accept this latter tenet, but would question that it has anything to do with cosmology. Verticality refers to an inward dimension, they would argue; it refers to something spiritual with which cosmology has no concern. But the matter is not quite so simple—nor quite so Cartesian, in fact. The inner and the external, it turns out, are profoundly related. I repeat: As our universe flattens, so does our culture. As Huston Smith points out: "A meaningful life is not finally possible in a meaningless world."

We stand in need of a new cosmology: of a cosmos incomparably more vast than the universe of contemporary astrophysics. I am not of course referring to spatial extension: the physical universe as currently conceived encompasses light-years enough. I speak rather of things which cannot be measured or weighed, of things, in fact, which can only be spoken of in traditional terms: of an integral cosmos made up of distinct ontologic levels, which we may picture as so many horizontal planes or concentric spheres.[7] I speak thus of a cosmic hierarchy, a universe with an added dimension: the dimension of verticality, which has to do, not with spatial direction, but with value and meaning, and ultimately, with first origins and last ends. It is the dimension that transforms the cosmos from a mere *thing* into a *bona fide* symbol: into a theophany, no less; it is thus the dimension that nourishes the artist, the poet, and the mystic in us—the dimension, as I have said, which enables us to be fully human. It is also, however, the dimension that in fact permits existence, permits being as such; for it can indeed be said, ontologically speaking, that nothing can exist simply on a horizontal plane. Nothing therefore can be understood, known, *experienced*—without entering into the vertical dimension. It is no small disadvantage, thus, that verticality has been banished: postmodernist nihilism, it turns out, is by no means unjustified. In fact, it is profound. Nietzsche was right: "We have abolished the true world. What has remained? The apparent one perhaps? Oh, no! With the true world we have also abolished the apparent one." Prophetic words!

But the question remains whether the true world can be reinstated. To be precise: Can hard science sanction a multi-level cosmology of the traditional kind? I submit that it can. What is more, I contend that science today not only permits a hierarchic cosmology, but indeed demands a worldview of this kind; it is only that science itself cannot articulate that demand. As we have seen in light of the perennial ontology, quantum mechanics itself entails three distinct ontological levels: the physical, the corporeal, and the intermediary. Turning to Dembski's theory, one finds that the newly-discovered criterion of design permits us to distinguish scientifically between the animate and inanimate levels of corporeal being, in accordance with traditional cosmology. To this hierarchic structure now comprising four tiers, Gibson's "ecological" theory of perception adds a further division through its discernment of perception as an act *sui generis*, irreducible to physiology. It thus appears that in its own way the theory distinguishes between the plant and animal levels within the biosphere. One thus recovers four links of what Arthur Lovejoy terms "the great chain of being," with a fifth link, comprised of the physical domain, tacked to the lower end. It is evident, to be sure, that our scientific grip on reality loosens the higher we ascend along this chain; and this is to be expected, given that our sciences are perinoetic, and that we are ascending towards the essential pole. The wonder is that modern science has been able to grasp enough of supra-physical reality to entail the aforesaid ontological distinctions. In accomplishing this task, however, it has reached the end of its tether; it is a science, after all, based upon physics, and geared, therefore, to deal precisely with the quantitative aspects of reality, which emanate, so to speak, from the material pole of existence, and do not extend past the corporeal domain. We should recall that the ontological discoveries which I have enumerated derive in fact from *negative* results: from the discovery of boundaries which the science in question cannot cross. These recognitions constitute what I have termed limit theorems; they are scientific findings, which reveal their ontological implications only in light of traditional doctrine. It is in this sense, then, that contemporary science entails a distinction between the physical, corporeal and intermediary degrees, as well as a subdivision of the corporeal corresponding to the traditional "mineral, plant, animal" trichotomy.

To proceed further, beyond such limit theorems, one requires means of a very different kind—which is, of course, exactly what the limit theorems themselves imply. Above the physical domain, *essence* comes into play, and this is what a science based upon the discernment of quantity cannot comprehend. The only way to grasp essence is through an act of perception, be it sensory or intellective; and this constitutes authentic knowing: a knowing based upon the fact that, in point of essence, the external world pre-exists in the human microcosm. As I have noted in the Introduction, we are able to know or "enter" a given level of cosmic reality only by actualizing the corresponding state in ourselves. And this, to be sure, has ever been the *modus operandi* of traditional science, which is primarily concerned, not with quantitative parameters, but precisely with essences. I might remark, moreover, that the conception of a science based upon "seeing" has not altogether disappeared from the contemporary West. Such a science was in fact championed by Johann Wolfgang von Goethe during the heyday of the Enlightenment; and while that science was for the most part derided and ignored for more than a century, it has now been rediscovered and is being investigated and applied in various domains. I have already mentioned the name of Henry Bortoft, the physicist whose book *The Wholeness of Nature* constitutes a remarkable tribute to the depth and range of that long-neglected science; significantly, the book is subtitled "Goethe's Way toward a Science of Conscious Participation in Nature." Whether it be in Goethean fashion or otherwise, I would only add that "conscious participation in Nature" constitutes in fact the *only* way of grasping cosmic reality in its essence; it needs however to be understood that we "consciously participate in Nature" precisely by going into ourselves:

into the depths of our own being. As I have said before, the key to the problem of knowing lies in the Delphic injunction "Know thyself"; but of course, here we are speaking of an essential knowing, which has little in common with the perinoetic knowing attainable by way of the Baconian enterprise.

One sees that contemporary science, stringently limited though it be in what it is able to know, endorses nonetheless a hierarchic cosmology of the traditional kind, and allows alternative modes of approach. A restoration of cosmology, unthinkable a century ago, has thus become theoretically feasible. Since the Enlightenment, Western man has lived intellectually in a flattened cosmos, a truncated universe of mere particles, persuaded that science had so decreed; and now one knows—or ought to know!—that we have been deceived. It was scientism, it turns out, that perpetrated the fraud; and this we know on the authority of science itself.

What is above all needed today is what Seyyed Hossein Nasr terms a "revival of tradition."[8] We have lost what Nasr calls "the sapiential dimension which lies at the heart of tradition":

> Already in the nineteenth century, what remained of knowledge of an originally sacred character had become more or less reduced [in the West] to either occultism or a purely theoretical philosophy divorced from the possibility of realization, while even as theory it remained incomplete.[9]

It turns out, however, that a revival of sapiential tradition has in fact begun, and that the resultant knowledge is already gaining ground in the hearts and minds of many; as Professor Nasr explains:

> The principle of cosmic compensation has brought to the fore the quest for the rediscovery of the sacred during the very period which the heralds of modernism had predicted to be the final phase of the depletion of human culture of its sacred content, the period whose dawn Nietzsche had declared a century ago when he spoke of the 'death of God'.[10]

In the third of his Gifford Lectures, Nasr chronicles this revival with copious references, a lecture to which I refer the interested reader. I will only note that this revival, remarkably, comes at a time when modern science itself, by the force of its own inner logic, calls for the restitution of the traditional worldview. We seem in a sense to have come full circle. This does not mean, of course, that contemporary civilization is ready to embrace the ancient wisdom: nothing, obviously, could be further from the truth. There can be no doubt that the present order, which in fact is a disorder, must pass away before such a rebirth can actually take place; and surely we know not how and when this shall occur. Yet in a way we have already come full circle, as I have said. Phases of history, or perhaps even *yugas* and *kalpas* in the Vedantic sense, do not simply succeed one another or alternate, but seem in a sense to overlap in their extremities and interpenetrate. One is reminded of the words of Christ when He declared: "The time is fulfilled, and the kingdom of God is at hand" (Mark 1:15), at a time when the old world, historically speaking, had yet millennia left to go.[11]

It may seem that by alluding thus to Christian eschatology, I am obscuring the issue; yet I surmise that the rebirth of sacred tradition in its sapiential modes does in fact have an eschatological significance, hard as it is for us to comprehend these things. Suffice it to say that the significance of that revival, which has begun in our time, is greater by far than we are normally disposed to believe, even as a single spark of sacred truth outweighs all the things of this world combined. The vistas to which modern science, in particular, now gives access in principle, by virtue of its limit theorems, extend indeed to the very portals of the world to come.

Notes

1. For instance, Michael Denton, *Evolution: A Theory in Crisis* (Bethesda, Md.: Adler & Adler, 1986); Phillip E. Johnson, *Darwin on Trial* (Downers Grove, IL.: InterVarsity, 1993); and Michael J. Behe, *Darwin's Black Box* (New York: The Free Press, 1996).
2. It needs to be noted that scientific facts may indeed be "theory laden," as philosophers of science have come to recognize. Yet, even so, a distinction is to be made between such facts and what I have termed scientistic fictions. The point is that scientific facts are not to be conceived in isolation, but presuppose the theoretical structure within which they are defined. The truths of physical science, at any rate, are captured, not in isolated facts, but in entire structures. What I have termed scientistic fictions, on the other hand, tend to be isolated and *ad hoc* tenets, unsupported by any testable structure. Consider, for example, the tenet of bifurcation: the fact that physics as such can be interpreted in non-bifurcationist terms (as I have shown in *The Quantum Enigma*) proves that this claim can derive no support from any physical finding. For other examples of major scientistic articles of belief I refer to my book *Cosmos and Transcendence* (Peru, IL: Sherwood Sugden, 1984), where I have dealt with these questions at some length.
3. See *The Ecological Approach to Visual Perception* (Hillsdale, NJ: Lawrence Erlbaum, 1986).
4. Ibid., 60-61.
5. Ibid., 31.
6. I have elucidated this contention in *Cosmos and Transcendence,* op. cit., Chapter 7.
7. An excellent account of hierarchic cosmology has been given by Seyyed Hossein Nasr in his 1981 Gifford Lectures. See *Knowledge and the Sacred* (New York: Crossroad, 1981), which also gives extensive references.
8. *Knowledge and the Sacred,* op. cit., Chapter 3.
9. Ibid., 94.
10. Ibid., 93.
11. On this subject I refer to my article "Eschatological Imminence and Biblical Inerrancy," *Homiletic and Pastoral Review*, March 1987, 21-27.

Reply to Wolfgang Smith*

by Seyyed Hossein Nasr

Professor Wolfgang Smith is one of the very few scientists who have devoted their lives to the pursuit of science and are at the same time rooted in the teachings of the perennial philosophy. Therefore his essay "*Sophia Perennis* and Modern Science" is very pertinent and reveals the essential truth that the tenets of perennial philosophy are not only significant in the domains of religious studies, traditional art, psychology, and the like. Perennial philosophy is also of the greatest importance for a revaluation of the philosophy of the modern sciences and for providing a meaningful framework for the understanding of these sciences and especially what constitutes their basis, namely, quantum mechanics. This latter point has been already treated in Smith's remarkable opus, *The Quantum Enigma*, and in fact much of his discussion in this essay is related to the theses of that work, although the present essay is an exceptional synthesis of Smith's thought on the subject of the relation between the perennial philosophy or *sophia perennis* and modern science and not simply a summary of *The Quantum Enigma*.

I am in such deep agreement with nearly everything written in this essay that there is hardly a point which I would wish to criticize. My response on most issues will be in fact simple confirmations. Nevertheless, there are a number of points upon which I wish to expand in order to clarify further my own views on the subject. Smith writes that "science in the traditional sense is thus a matter of 'reading the icon'—a far cry indeed from the Baconian vision!" This is a very apt manner of speaking of the traditional cosmological sciences. These sciences depict a cosmos which revealed a meaning beyond itself and can be contemplated as an icon. One could in fact go a step further and say that, using the language of Christianity, the cosmos *is* an icon and can only be understood in depth as an icon, which reveals a divine reality beyond itself. The traditional cosmological sciences brought out this iconic reality and permitted those who studied and understood them to *see* the cosmos as an icon and to be able to contemplate it rather than knowing it only discursively. The modern sciences, issuing from what Smith calls "the Baconian vision," also know nature but no longer as an icon. They are able to tell us about the size, weight, and shape of the icon and even the composition of the various colors of paint used in painting it, but they can tell us nothing of its *meaning* in reference to a reality beyond itself. What they tell us about the size, composition of the

* The essay to which this refers is Chapter 1 of this book.

paints, and so on of the icon are not false on their own level, but they do not exhaust knowledge of the icon and it would be both ignorance and *hubris* to claim that this type of knowledge is the only knowledge possible of the icon. The consequences of this ignorance parading as totalitarian science combined with *hubris*, made even more dangerous by being denied, are lethal for man's spiritual life. This ignorance, *hubris*, and denial have dire consequences even more outwardly in man's relation to the world of nature, divorced in the modern world from ultimate meaning as a result of the exclusive claims of Baconian science.

Smith qualifies his comments about Baconian science by saying that contemporary science at its best is not completely Baconian as Einstein's occasional comment about the "Old One" suggesting that he too may have been searching for "*vestigia* of a kind" shows. I agree that there are a number of individual scientists, even in the contemporary period, who, like the English botanist John Ray, cultivated science with the goal of discovering the *vestigia Dei* in creation. But they were and are functioning within a scientific framework in which such concerns by an individual scientist could not in any way affect the science they have produced. One can study and accept the theory of relativity with or without references to the "Old One," which means that the result of Einsteinian science is not the search for the *vestigia Dei* and the depiction of the cosmos as an icon, whatever may have been Einstein's personal views and attitudes.

The distinction that Smith makes between what *ought* to transpire in the mind of a modern scientist, which is reasoning upon data provided by the senses, and the meaning of knowledge according to the *sophia perennis* is of crucial importance. The epistemology provided by the *sophia perennis* covers "an incomparably greater range of cognitive possibilities" to quote the author, since it relates all acts of knowing to participation of the human intellect in the light of the Divine Intellect. In much of the discussion going on today in the domain of epistemology, both scientific and otherwise, this issue is forgotten or at least not emphasized, including the works of many contemporary Muslim thinkers writing on the subject. This participation is not confined to "moments of illumination" but involves all knowledge which relates the human subject to the object that is known. Smith quotes the Gospel of John and states, "what ultimately connects the human subject to its object in the act of knowing is indeed 'the true Light which lighteth every man that cometh into the world'."(1:9) Since the *sophia perennis* is both perennial and universal, one needs to add here how central this thesis is in other traditions. Being better acquainted with the Islamic tradition than with others, I can turn to that tradition and add that there are numerous verses of the Quran and *Hadîth* about the relation of knowledge (*al-'ilm*) to light which constitutes a vast hierarchy issuing from God. In fact according to the Quran (XXIV:35) God is the light of not only the heavens but also the earth. On the basis of these traditional sources and certain elements of Greek philosophy, Islamic philosophers, going back to al-Fârâbi and Ibn Sînâ, spoke of the illumination of the human intellect by the Active Intellect in the act of intellection. The symbolism of light was particularly central to the teachings of Suhrawardî, the founder of the School of Illumination (*al-ishrâq*) and what Professor Smith has written on participation in the Light of the Divine Intellect is practically identical with the views of Suhrawardî, except that the latter identifies this light in its various degrees with the different angelic substances. Again I need to emphasize how important this issue is for an in-depth study of the epistemology of modern science in light of traditional teachings, and as far as Islamic thinkers now writing on the subject of epistemology are concerned, how crucial it is first to understand the tenets of traditional Islamic philosophy in this domain before embarking upon often puerile and moreover fruitless comparisons of traditional epistemologies and modern scientific epistemology.

Smith points quite correctly to the need for a living sapiental tradition within which cosmological sciences are cultivated and without which they wither and die "thus giv[ing] rise to what may indeed be termed a superstition." This is a very correct assessment upon which I must elaborate. A superstition is literally something whose ground has been removed. The metaphysical teachings of the *sophia perennis* constitute precisely the ground upon which the traditional cosmological sciences stood. Therefore, with the destruction of that ground these sciences could not but be reduced to superstition, although they still carried residues of truths no longer understood. The whole phenomenon of occultism in the West is quite interesting from this point of view. In other civilizations where the metaphysical ground has not been destroyed, there are certainly forms of popular superstition which are also very much present in the modern world, albeit in other guises, but in those civilizations one does not encounter the phenomenon of occultism as it developed in the salons of France and elsewhere in Europe from the eighteenth century onward. Many modern people, especially those of a scientific bent, immediately dismiss the traditional sciences such as alchemy as being simply superstition. But they do not realize that such sciences are like jewels which glow in the presence of the light of a living sapiental tradition and become opaque once that light disappears. Paradoxically enough, by claiming to relegate the traditional cosmological sciences simply to the category of superstition, the modern scientific enterprise has not only been helpless before the mushrooming of interest in these traditional sciences even in their residual form known in occultist circles, but it has been instrumental in the rise of new forms of superstition, such as the idea of progress, which are much more dangerous for the future of humanity than the practice of predictive astrology.

Returning to the question of perception, Smith emphasizes that light waves, sound waves, brain function, and the like are of course necessary and play a role in the act of perception but that "they do not—they cannot!—constitute the perceptual act" which is "literally not of this world." He also lays the error of belief in bifurcation and the lure of it at the feet of the state of forgetfulness of "participation" by post-medieval European man. I have restated these lines to emphasize their central importance and my full agreement with it. Modern philosophy, psychology, or science are simply not able to explain perception which they always reduce to one of its parts or something else because the participation of the human intellect in the Light of the Divine Intellect is simply beyond the truncated worldview within which all modern thought, whether it be philosophical, psychological, or scientific operates. The rediscovery of the real significance of perception is only possible in light of the *sophia perennis* and is itself a key for the discovery of the metaphysical universe depicted by the perennial philosophy in its vastness and wholeness. And I agree with Smith that the greatest obstacle to the integration of modern science into the higher orders of knowledge and the rediscovery of how the miracle of perception works is Cartesian dualism or the theory of bifurcation.

I believe that the ingenious distinction made by Smith between the corporeal and physical worlds and the confining of quantum mechanics to the physical rather than the corporeal world, as well as the relation between the two of which he speaks, are major steps in the formulation of a more meaningful philosophy of physics in accordance with the *sophia perennis*. Like Smith, I also wish to emphasize that the corporeal and the physical worlds, as defined by him, are not only different but constitute two ontologically distinct domains. There is an ontological hiatus between the two and one cannot say that this physical stratum contains simply the "building blocks" for the corporeal world. The whole idea of fundamental particles from which we can build up the corporeal world with its forms and qualities is therefore false; and form and quality associated with the corporeal world—not to speak of psychological and spiritual realities—can never be reduced to the quantitative elements of the physical world which alone can be studied through the "modern scientific method" and be made to constitute the subject of quantum mechanics. Naturally, without mathematics there is no possibility of study of that physical world, as defined by Smith.

To understand the ontological status of the corporeal and physical worlds would also solve the status of quantum physics not only in its relation to the world of classical physics but also to the corporeal, or what is ordinarily called the "physical world" (contrary to Smith's terminology), and the traditional sciences which deal with the qualitative and formal aspects of that world. This ontological awareness would also make clear the basic point I have mentioned in so many of my writings, namely, that the traditional cosmological sciences are not simply crude attempts to understand nature and primitive stages of the modern quantitative sciences, but contain profound knowledge of the formal and qualitative aspects of the corporeal world not reducible to quantity nor of lower significance than quantity *à la* Galileo. On the contrary they refer to realities with a higher ontological status than the quantitative.

The analogy made by Smith between the *materia secunda* and the Euclidean plane from which geometric forms emerge and in which they become actualized is also a brilliant one with which I could not but agree. It is so important to remember that the objects of modern physics are not like tables and chairs except smaller, but that they are "constructed" to quote Smith; that is, "they are defined or specified by a certain experimental intervention." And yet, although experimental means affect the form of the physical laws, the content of the mathematical equations which contain the laws do not derive from the particular experimental methods used. Here again the analogy between the *materia secunda* of physics and the geometric plane of Euclidean geometry becomes useful, because in the case of geometry also, while the manner of operation of a geometer affects the geometric properties of what is constructed, whether the geometer draws a square or a circle, the mathematical structures of the geometric forms are not determined by the geometer. They belong ultimately to the intelligible world which Smith quite rightly associates with the Platonic understanding of this term. This whole analysis is very much in accord with my views and opens the door for those who are able to understand the tenets of the *sophia perennis* as well as modern physics, for the integration of quantum mechanics into the traditional hierarchy of knowledge. It invites the integration of the world with which quantum mechanics deals into the ontological hierarchy of the perennial philosophy.

To achieve this end, it is of crucial importance to realize that the physical objects of quantum mechanics are not "things" in the ordinary sense but of much smaller dimension. Rather, they are relational and do not possess an *esse*. The main lesson to learn from this truth is that the essential attributes of things, therefore, come not from the quantum objects but from elsewhere or more precisely from above. Smith offers a wonderful metaphor by saying, "the essence of a plant, after all, derives from the seed, and not from the ground in which the seed is planted." One might say that the seed in this metaphor refers ultimately to the archetypal reality or Platonic Idea which the particular plant in question reflects and manifests on the corporeal plane. It is a valuable contribution by Smith to assert that modern physics does not deal with *essence*, that it is "*inessential*," because the realization of this truth provides an opportunity for those reductionists who are looking for the sun in the bottom of the well to cast their eyes above in order to see the origin of the essences which one observes and experiences in the corporeal world.

From my perspective, Professor Smith is completely right when he asserts that "the decisive step in the restitution of the *cosmologia perennis* is without question the rediscovery of 'forms' as an ontological and causal principle," and in fact on several occasions in my writings I have expressed the same idea. And it is also completely true that the destruction of the significance of forms based on incomprehension of their real significance by Bacon and Descartes opened the door for the reductionism of modern science and the constant attempt by scientists to explain things by their parts, asserting that the whole is no more than the sum of its parts. When Smith states, however, that this destruction of forms "has not . . . led towards the realization of the reductionist goal" and that "the reductionist philosophy appears also to have outlived its established usefulness as a heuristic principle," it seems to me that he is looking only at a few scientists like himself and not at the impact of reductionism associated with science on the general cultural scene of today. It is enough to look at the current mainstream view of the physical world, of medicine and the body, of the approach to the solution of social, economic, and ecological problems to see how entrenched the reductionist view really is. The philosophy of wholeness still remains in the margin of modern and postmodern man's worldview thanks mostly to mainstream modern science.

Furthermore, Smith makes the crucial statement that there exist in the world non-mathematical formal principles which are "none other than the aforementioned substantial forms, which prove moreover to be 'essential' in a strict ontological sense." He adds further that these forms constitute an ontological order. If this central statement of Smith's, with which I agree whole-heartedly, were to be accepted fully by modern science, a new scientific view would be born that would cease to be that of modern science as it is known today and become transformed into a further extension of the traditional sciences as I have already proposed. Then and only then could one say that reductionism has ceased to be operative and has outlived its "usefulness." Until then, unfortunately reductionism continues to level things to their lowest common denominator, to destroy quality in the name of quantity and to impoverish the spiritual vision and the minds of those affected by its siren call.

The essay of Wolfgang Smith, to which I have only added a few commentaries, is a seminal essay and should be studied carefully by all interested in the reintegration of science into the metaphysics contained in the heart of the *sophia perennis* and the traditional cosmological sciences associated with it. It should also be necessary reading for all searching for a new and richer philosophy for science, for those who often end with superficial adaptations of Taoism or Hinduism which are then related to the findings of modern science. Smith takes us much further in this quest and shows the role that the *sophia perennis* can play in the veritable understanding of the significance of quantum mechanics and the integration of scientific knowledge into the universal hierarchy of knowledge.

For some forty years I have been writing on traditional science in relation to modern science, on Islamic science, and the hierarchy of knowledge as well as on traditional metaphysics and the perennial philosophy. It is an exhilarating experience for me to see here almost a synthesis of my own thought with many new insights on the subject presented by an active scientist also well versed in traditional doctrines. In writing these lines I feel as if I am expanding some of my thoughts in the direction of a new horizon opened by Smith. There is practically nothing in this important text to which I would need to respond in a critical manner in order to clarify differences with my own thought. On the contrary to understand my thought on the subject of the relation between the *sophia perennis* and modern science, it is important to pay attention to my confirmation of the main theses of Smith's essay which do full justice on the one hand to traditional doctrines and the tenets of the *sophia perennis*, to quote his terminology, and on the other to the discoveries of quantum mechanics and the nature of the whole venture of modern physics. I am happy that the occasion to write for this volume made it possible for Professor Smith to produce this exceptionally important essay in the field of the relation between the perennial philosophy and modern science, a field which has preoccupied me since my student days.

Acknowledgements

The essays published in this collection have been reprinted by permission from the publishers as follows:

Chapter I Reprinted (in abridged form) by permission from *The Philosophy of Seyyed Hossein Nasr*, edited by Lewis Edwin Hahn, Randall E. Auxier, and Lucian W. Stone, Jr. (La Salle, IL: Open Court Publishing Company, 2001), copyright © 2001 by The Library of Living Philosophers.

Chapter II First presented as a Templeton Lecture on Christianity and the Natural Sciences at Gonzaga University in February 1998, and delivered a second time before the 1998 Thomistic Institute at the University of Notre Dame. Reprinted with minor changes from *The Thomist* 63 (1999).

Chapter III Reprinted (in revised form) from *Sophia*, Vol. 6, No. 2 (Foundation for Traditional Studies: Oakton, VA, 2000).

Chapter IV Reprinted from *Sophia*, Vol. 3, No. 1 (1997).

Chapter V Reprinted (with added material) from *Sophia*, Vol. 5, No. 1 (1999).

Chapter VI Reprinted from *Sophia*, Vol. 6, No. 1 (2000).

Chapter VII Reprinted (in revised form) from *Sophia*, Vol. 7, No. 2 (2001).

Chapter VIII Reprinted from *Sacred Web*, Vol. 9 (2002).

Chapter IX Reprinted from *Sophia*, Vol. 8, No. 1 (2002).

Chapter X Reprinted from *Sacred Web*, Vol. 10 (2002).

Chapter XII Reprinted (in revised form) from *Modern Age*, Vol. 43, No. 1 (2001).

Appendix Reprinted by permission from *The Philosophy of Seyyed Hossein Nasr*, edited by Lewis Edwin Hahn, Randall E. Auxier, and Lucian W. Stone, Jr.

Index

CPSIA information can be obtained
at www.ICGtesting.com
Printed in the USA
FSOW01n1038010817
37108FS

9 780962 998478